Henri Lefebvre

Henri Lefebvre

A Critical Introduction

ANDY MERRIFIELD

Published in 2006 by
Routledge
Taylor & Francis Group
270 Madison Avenue
New York, NY 10016

Published in Great Britain by
Routledge
Taylor & Francis Group
2 Park Square
Milton Park, Abingdon
Oxon OX14 4RN

Routledge is an imprint of Taylor & Francis Group

Printed in the United States of America on acid-free paper
10 9 8 7 6 5 4 3 2 1

International Standard Book Number-10: 0-415-95207-7 (Hardcover) 0-415-95208-5 (Softcover)
International Standard Book Number-13: 978-0-415-95207-1 (Hardcover) 978-0-415-95208-8 (Softcover)
Library of Congress Card Number 2005024643

Library of Congress Cataloging-in-Publication Data

Merrifield, Andy.
Henri Lefebvre : a critical introduction / by Andy Merrifield.
p. cm.
Includes bibliographical references and index.
ISBN 0-415-95207-7 (hb : alk. paper) -- ISBN 0-415-95208-5 (pb : alk. paper)
1. Lefebvre, Henri, 1905- 2. Social scientists--Biography. 3. Social sciences--Philosophy. I. Title.

H59.L44M46 2006
300.92--dc22 2005024643

Visit the Taylor & Francis Web site at
http://www.taylorandfrancis.com

and the Routledge Web site at
http://www.routledge-ny.com

À Corinna, avec une jeunesse du cœur …

Over the future, everybody deludes themselves. We can only be sure of the present moment. Yet is this true? Can we really know the present? Are we capable of judging it? How can somebody who doesn't know the future make any sense of the present? If we don't know towards which future the present leads us, how can we say that this present is good or bad, that it merits our support or our mistrust, or even our hatred?

—Milan Kundera, *L'ignorance*

Contents

Foreword: Something Cool

The city of psychoanalysis salutes the land where the Children of Marx and Coca-Cola grew up, or in many cases refused to. But I'm showing my age. It's been many years since New York could muster up an audience for angst and Woody Allen. And Godard's cool Left Bank cocktail of student disaffection and *ye-ye* nonchalance lost its sparkle long before Brigitte Bardot started flirting with M. Le Pen.

Moreover, relations between these two places have become more strained even as globalization has thrown everyone into everyone else's laps. As I write this, the United States is still in denial that its position in the world is rapidly approaching the shrunken irrelevancy of post-imperial Great Britain. We're just an island, separated from the great land mass of Eurasia, pretending that we own the joint, indulging expensive fantasies about enlightening the world while turning our backs on countries where concepts of Enlightenment took root. Not good.

Manhattan, meanwhile, is an island within an island, imagining itself as a leader of Blue State (liberal, Democratic) sensibilities while the record strongly indicates that it has actually been the reverse. It is New York that has given us Rudolph Giuliani's chronic hostility toward the First Amendment; fake pre-modern architecture; news magazines featuring cover stories on assorted mythological religious figures; and other convulsions of Red State backlash against the long-ago 1960s and in general anything that breathes.

"Haven't they heard Marx is dead?," Giuliani replied when asked about his plans to uproot community gardens to make room for luxury housing. Heave-ho! While *The New York Review of Books* canceled its subscription to liberal thought by publishing stories that claimed to discredit Freud.

But never mind. So long as Andy Merrifield is living in France and I remain in New York, Marx and Freud are still conversing with one another across the divide of water, ideology, and time. In my perplexed head, Merrifield and I are two halves of a whole. We're one of the last great Surrealists. Our mutual *amour fou* is the city: the courting ground of crazy love.

Merrifield's got his Marx down better than I'll ever get my Freud. He's become a one-man, year-round, world-wide festival of non-vulgar Marxism. I'm just an old-fashioned nut case. But Surrealists are not required to be experts in either field. We define our own field. It is bordered on two sides by history: one border chronicles the outer life, the other one tracks the inner. Mr. M and Mr. F sit like referees atop tall spindly chairs on either line. Out! In!

The remaining sides of the field are left open. One of these trails off into the future, the other drifts backward toward a ground of origin that is unknowable apart from myth. Culture happens in the middle. The forms it takes are not invariably symptoms. But sometimes the symptoms are thrilling also. We set up bleachers. We boo and cheer.

* * *

Merrifield booed me once. I didn't mind too much. When smart people boo, you should take it as an invitation to call them right up and say, "Hi, there!" Merrifield had taken offense at a reference I made to Guy Debord in a story about Times Square. I seem to recall his saying something mild, like "Debord would have hated Muschamp and everything he stands for." Setting aside the rhetorical propriety of airing one's sentiments through the mouths of dead people, I nonetheless decided to interpret the boo as an instance of what Buddhists call "negative attachment." The attachment is the main thing. The negativity is the attachment's shadow. And it doesn't pay to get too caught up with shadows. Often, they're just there to be enjoyed, like any play of light on the wall. Sometimes they boo. Sometimes they act scarier and say "Boo!" But the smarter breed just wants to come in and play.

The city that we hate is also the city that we love. It strikes me that Henri Lefevbre's work and Andy Merrifield's both spring from this variation on what Melanie Klein called the depressive position. It is the business of a city to offer something for everyone to hate, even to present itself as completely hateful to some people most of the time. But even Frank Lloyd Wright, who devoted endless energy to denouncing the city as "the Moloch that knows no god but More," couldn't resist being swallowed up by New York from time to time. And the intensity of his attachment was even more evident in the passion of his attacks. I share with Sybil Maholy-Nagy the view that the city is the matrix of man, whatever our feelings might be toward it. And try as you may you simply cannot keep Mother down in the fruit cellar forever. She will not stay there.

Henri Lefebvre introduced a stance of radical ambivalence toward the city in his book *The Urban Revolution*, first published in France in 1970 but not issued in English translation until 2003.

He dramatizes this position in his first chapter, "From the City to Urban Society," with a dialectical exchange for and against the concepts of streets and monuments. The likelihood is that most readers can identify with both positions, in whole or in part. Neither is unintelligent, and the recognition that we can and do live with these opposites is tonic. We can defer judgment until we're more adept at grasping the dialectic process revealed by the unfolding of urban experience.

Freudians and Marxists are similarly engaged in the dispelling of illusion, however, and as a journalist I am naturally interested in all techniques that have shown themselves effective in performing this task. Journalists are also supposed to be involved in this, but the truth is that we create, perpetuate, and reinforce at least as many illusions as we dispel. But the ideal is not dead, and if it is not always possible for us to draw aside as many veils as we might like for readers—because, among other reasons, we want to share with them the delight we feel when an illusion is done well—we can at least point them in the direction of writers like Andy Merrifield who can accompany them further on the road to the enlightenment that even the most misguided among us are actually seeking. The Buddhists call this unveiling process *shakubuku*. I call it Shake and Bake.

* * *

Lefebvre pulled the plug on formalism: that was his decisive contribution to those who regard buildings primarily as pieces of the city, not as autonomous works of art. I hasten to add that this disconnection represented an expansion of aesthetic values, not a denial of them. What Lefebvre rescinded was the equation of aesthetics with the simplistic brand of formalism promoted by the Museum of Modern Art. Philip Johnson, who played an important role in adapting that brand for architecture, used to chide me because my writing "isn't about architecture, of course." The last two words carried the burden of Johnson's meaning. The subject was closed to discussion. Architecture was the moving and shaping of geometric forms in two and three dimensions. All else was sociology, a waste of time. Nice work if you can get it.

And there had been a time when formalism was a radical position, and to take it was to embrace a broad set of progressive causes. And it is still available both as an analytic tool and as an episode in the history of taste. I take to heart Roland Barthes's warning that the enemies of formalism are "our enemies," they are the people who claim the authority to enforce a strict correspondence between signs and meanings. Unfortunately, by the 1960s, the enemies had become very shrewd in manipulating forms to cleanse the images of toxic enterprises. *The Life of Forms in Art* had become the Death of Art in Logos.

We are workers, producing our own factory just by walking down the street: that's one way to summarize what I took away from Lefebvre's *The Production of Space.* And I say this as a former window dresser, who once had the good fortune to work inside one of those precious glass-enclosed storefront stages. Store display was the only form of design I ever worked in, and I loved it: the commercialism as well as the aesthetics (the store, which no longer exists, specialized in Good Design objects) and the effect that the fusion of commerce and appearances can have on the life of the street. It was like being a sidewalk painter, and if people believed that buying the objects on display enabled them to acquire the image I had set up for their viewing pleasure—if the window got enough of them to cross the threshold into the shop—that was good enough reason to ask for a raise.

But a sidewalk painter isn't only or even mainly working for paying customers, and neither was I. I was working for passers-by, for the wonderful ladies who get all dressed up to go out window-shopping, almost as if they were going to the opera, and for couples staggering romantically around the city after midnight, and for gentlemen out cruising (with or without dogs), and for everyone who appreciates the seductive pleasure of seeing their reflections in the glass and in the temptations behind it.

Try seeing things from a window-dresser's point of view. For us, the sidewalk is the stage, the people walking along it are the players, and we are the audience for a live version of that wonderful *Twilight Zone* episode where each of the mannequins in a department store gets to live for a day among the world of shoppers. Remember? And how one of the mannequins forgot that she wasn't human and had to go back in the window at the end of her special day? The solitary *flâneur* is also a spectacle—not just the beholder of them.

To a Mahayana Buddhist, the fusion of city and self is more than a poetic metaphor. It illustrates a concept called *esho funi*, roughly translated as the oneness of life and its environment. Literally "two but not two," they are different aspects of the same entity, like the heads and tails of a coin. In his depiction of urban space, Lefebvre has taken what strikes me as a Western route toward a similar concept. It is not mystical, but then, to a Mahayana Buddhist, neither is *esho funi*. "Two but not two" is a construction that represents *things as they are*. And the goal of Buddhist practice is to bring one's subjective perception into closer alignment with things as they are. That is what Enlightenment means. Lefebvre's arguments against subjectivism are thus enlightened in both the Eastern and Western senses of the term.

* * *

I hope that Andy won't be offended if I propose him for honorary membership in the illustrious guild of window-dressers. Arguments about the contemporary city are his wares: the ideas and the thinkers who articulated them have already inspired

Merrifield's vibrant writings on Guy Debord, Walter Benjamin, David Harvey, and other students of the great human matrix. He may tempt you to cross the threshold to pursue those authors at closer range. If you've already read them, he will recast their thoughts in the lively light of his own imagination.

Merrifield's contribution to the literature on cities is substantial in its own right. It reflects the transformation of the urban public into a fluid and complex social arrangement of audiences: groups of individuals organized for the purpose of obtaining information to which they might be unable to gain access if they were acting on their own. The information might take the form of a symphony concert, a website, or simply the experience of rubbing shoulders together in a crowded place; it might be found between the covers of a book or on a computer screen, scrolling through reviews by a book's readers.

It takes a great audience to make a great performance: it takes the multiple massing of curiosity, receptivity, and a strong desire to share—qualities that vibrate throughout Merrifield's literary portraits. What we see through the lens of Merrifield's writing is the emergence of an audience for the city: people who are drawn to it by the desire to share sidewalks and shop windows with others similarly inclined. Like all great critics, Merrifield sharpens the audience's appreciation of the experience, as he also helps to define who we are and to show, in the process, that we actually do exist.

À nos amours.

Herbert Muschamp

Acknowledgments

I'd like to thank Neil Brenner and Marshall Berman for helping kick-start this project and Dave McBride at Routledge for keeping it typically on track. I owe David Harvey an enormous and incalculable debt for his intellectual guidance and friendship over the years and for opening my eyes to Henri Lefebvre; Ed Soja's Lefebvrian interventions always provoke and instruct, as do those of my pal Erik Swyngedouw. The long discussions with him and with Clive Barnett in Oxford's King's Arms, oh so long ago now, remain a vivid and cherished memory.

Preface: "A Youthfulness of Heart"

I never met Henri Lefebvre or saw him lecture. Some of my friends who did said he was a real knockout. Others who had contact with him recall his warm, slow, melodious voice, his boyish passions, his virility—even in old age—and the posse of young, attractive women invariably in his train. Portraits cast him as a Rabelaisian monk and Kierkegaardian seducer all rolled into one. I'm sorry I missed this act, missed the man himself, *en direct,* live. But I did see him on British TV once, back in the early 1990s. The series *The Spirit of Freedom* was strictly for insomniacs and appeared in the wee hours on Channel 4. Each of the four programs tried to assess the legacy of Left French intellectuals during the twentieth century. The cynical and pejorative tone throughout wasn't too surprising given that its narrator and brainchild was Bernard-Henri Lévy, France's pinup thinker and *Paris-Match*'s answer to Jean-Paul Sartre.[1] The night I watched, an old white-haired man sat in front of the camera, dressed in a blue denim work shirt and rumpled brown tweed jacket. In his ninetieth year, it was obvious to viewers Lefebvre hadn't long left to live. Even Lévy described his interviewee as "tired that afternoon. His face was pallid, his eyes blood-shot. I felt he was overwhelmed from the start and clearly bored at having to answer my questions."[2]

I didn't care that Lefebvre looked tired and bored that night. I remember he kept telling Lévy he'd rather talk about the present and the future, about things going on around him in the world, rather than recount tales of bygone days. More than anything, I'd been overjoyed to glimpse the old man himself, and I still vividly remember the moment. It was my first real sighting of a scholar who'd stirred my intellectual curiosity for several years already. The long-awaited English translation of *The Production of Space* had just appeared in bookstores around that time, and Lefebvre was much in vogue within my own discipline, geography. I was still in the throes of my doctoral thesis, too, using his work as theoretical sustenance; my first published article, bearing his name in its title, had been accepted in a professional journal. I felt like I was about to enter the adult world of academia with Lefebvre as my guiding spirit, a man I admired not just for what he wrote but for how he lived. His rich, long, adventurous life of thought and political engagement epitomized for me the very essence of an intellectual. I found him refreshingly different from the post-Sartrean "master thinkers" like Foucault, Derrida, and Althusser, more in touch with everyday life and everyday people; Lefebvre spoke to me as a radical person as well as a radical brain.

I loved his grand *style*. He wasn't afraid to think about politics and current affairs on a grand, sweeping scale or to philosophize what he called "the totality of life and thought." Lefebvre wanted to "de-scholarize philosophy," wanted to make it living and pungent, *normative* and holistic.[3] Indeed, "to think the totality" was Lefebvre's very own pocket definition of philosophy itself, the magic ingredient of his "metaphilosophy," through which, like

Marx, he'd grasp everything "at the root." It was his grand style that a *Le Monde* obituary emphasized when Lefebvre peacefully passed away, a few days after his ninetieth birthday, on June 29, 1991. The major daily described his life, as only the French could, as "adventures of a dialectician."[4] They bid adieu to the "last great classical philosopher," to the last great French Marxist, in a valediction that hailed the demise of not so much a generation as a mode of thinking. The obituary almost implied that Lefebvre's departure signaled the end of the twentieth century, the "short twentieth-century" that Eric Hobsbawm described, replete with all its promise and horrors.[5] Lefebvre knew that century first-hand. He traversed its big historical shifts and tumultuous events, its world wars, its major avant-garde movements. He'd belonged to the French Communist Party; fought against fascism for the Resistance Movement; lived though the growth of modern consumerism, the age of the Bomb, and the cold war; and witnessed the tumbling of the Berlin Wall. He'd driven a cab in Paris, broadcasted on radio in Toulouse, taught philosophy and sociology at numerous universities and high schools, and godfathered the 1968 generation of student rebels. Lefebvre was a man of action as well as ideas.

He was a Marxist who introduced into France a whole body of humanist Marxism. But he was a Marxist who seemed to reinvent himself, conceive a new sound, probe a new idea, reach a new note, almost every decade. Each reinvention built on an already accomplished body of work, yet took it further, propelled it onward. Frequently, these restless formulations recreated the old world in a new way; other times they somehow anticipated what was about to unfold in reality. He authored more than sixty books, since translated into thirty different languages, and made brilliant analyses on dialectics and alienation, everyday life and urbanism. The "retired" professor never let up in the 1970s and 1980s, never rested on his emeritus laurels. Always peripatetic, always inquisitive, he continued to travel far and wide, making prescient analyses on the changing nature of the state and the role of space in the "survival of capitalism." As an oeuvre, Lefebvre's fascinating breadth and imaginative reach are perhaps unmatched. Manuel Castells, a former assistant of Lefebvre's at Nanterre in the late 1960s, once remarked that Lefebvre "doesn't know anything about how the economy works, how technology works … but he had a genius for intuiting what really was happening. Almost like an artist … he was probably the greatest philosopher on cities we have had."[6]

Lefebvre blasted out his books "jerkily, hastily, nervously."[7] This modus operandi is the story of his whole literary life; it was a habit he'd never relinquish, whether in war or peace. He wrote every book as if it was his last: feverously, rapidly—perhaps, on occasion, too rapidly. Many, in fact, were dictated, the spoken word transcribed on the page by faithful secretaries, current girlfriends, or a latest wife. Arguably, he undertook too much during his long career, conceiving brilliant, original projects yet rarely completing any of them, leaving them instead gaping, incomplete, suggestive, as he flitted on to something else. "I loved too much," he admitted in his autobiography *La Somme et le Reste* (*Tome I,* p. 48), "the bubbling and the fermenting of an idea that burst out new and fresh." On the other hand, this is what made his work so experimental, so approachable; you can enter it and write your own chapter within it. Lefebvre's method followed Jean-Paul Sartre's ideal method: "It is the nature of an intellectual quest," Sartre said in *Search for a Method,* "to be undefined. To name it and to define it is to wrap it up and tie the knot."[8]

La Somme et le Reste is one of the most original works of Marxist philosophy ever and, to my mind, Lefebvre's greatest book. Written between 1957 and 1958 while Lefebvre worked in Paris at the Centre National de la Recherche Scientifique (CNRS), the two-volume, eight-hundred-leave tome was manically drafted each morning in an apartment along the rue de la Santé. He let everything rip, loosened every shackle. He was about to quit the Communist Party, to expel himself, departing from the left wing. (*J'ai quitté le Parti par la gauche,*" he enjoyed bragging.) Stalin's misdeeds were now public; the 1956 Soviet invasion of Hungary had disgusted many communists, Lefebvre included. This was Lefebvre's heart laid bare, his settling of accounts—with the party and with Stalinism—his "inventory" of the epoch: personal reminiscences and stinging rebuttals, historical and political analyses, literary set pieces, poems that hint of Rimbaud and Mallarmé, portraits of friends and pillories of enemies, all laced with dense philosophical disquisitions and Marxist delineations.

It was a lyrical and romantic "confession," revealing the struggles and delights of a life in philosophy as well as the pitfalls befalling a philosopher in life. The spirit of Rousseau seems close by; yet we also suspect Lefebvre remembers Dostoevsky's underground man's warning: vanity will always force men to fictionalize themselves. Few scholars nowadays could match Lefebvre's prosaic powers and grip on his times. Fewer still could ever dream up such an idiosyncratic book. Lefebvre is so passionately engaged with what was going on around him, and inside him, that, like the "Wedding-Guest" from Coleridge's "Ancient Mariner," we cannot choose but hear. Writing it was clearly therapeutic, a Proustian moment, reclaiming lost time and space: "This book," he wrote in 1973, in an updated preface, "speaks of deliverance, of happiness regained. Liberated from political pressure as one exits from a place of suffocation, a man starts to live, and to think. After a long, long period of asphyxiation, of delusion, of disappointments concealed … look at him: he crawls up from the abyss. Curious animal. He surges from the depths, surfaces, a little flattened by heavy pressures. He breathes in the sunshine, opens himself, displays himself, comes alive again."[9]

* * *

Lefebvre may have been the most self-effacing and least narrow-minded Marxist who ever lived, a utopian cognizant of the discredited utopias of the Eastern Bloc. A feisty critic of Stalinism from its inception, Lefebvre spent thirty years ducking, diving, and dodging the French Communist Party bigwigs, who followed orders from Moscow and took no prisoners. These endless run-ins with the hacks, and his rejection of Soviet-style socialism, never squared to a rejection of socialism, or of Marxism, because neither in the USSR bore any resemblance to Lefebvre's democratic vision. "Socialism until now," he claimed, "failed before the problem of the everyday. Too bad for it! It had promised to change life, but only did so superficially. Hence the profound dissatisfaction. … We always speak of economic deficiencies in socialist countries. But that isn't it. The wound there is that everything became too serious, horribly serious. They didn't know how to improve the everyday for real people … life was monotone, monochrome, tainted by a repetitive ideology."[10]

Lefebvre endures these days as a rare and necessary prototype: a utopian intellectual *engagé,* somebody who moved with the times yet helped shape and defy those times, interpreting the world at the same time as he somehow changed it. Philosopher cum sociologist, sociologist cum literary critic, literary critic cum urbanist, urbanist cum geographer, he was too eclectic to be any one of those categories alone. Too communist to be a romantic, too romantic to be a communist, his oeuvre bewilders and bedazzles, defies pigeonholing and classification, and makes a mockery of the disciplinary border patrols now stifling corporate universities, the University, Inc. Who could conceive Professor Henri fidgeting nervously at the next departmental research evaluation or getting the gripes about tenure when so much more is now at stake? "I am in essence," he stated in *La Somme et le Reste,* "oppositional, a heretic. … I pronounce myself irreducibly against the existing order … against a 'being' that searches for justifications beyond judgment. I think the role of thought is to harry what exists by critique, by irony, by satire. … I refuse to condemn spontaneity, that of the masses and that of the individual, even when it tends to be thoughtless, humorous, and bitterly ironic. I merit the value of spontaneity; life shouldn't fall from above and rest heavily; and everyday life and humanity aren't the realization of politics, morality, the state and Party."[11]

Lefebvre was a Marxist who sought not to denounce student exuberance in 1968 but to foster it, to use it productively, constructively, tactically, alongside skeptical working-class rank and filers. In *The Explosion* (1968), scribbled as the Molotov cocktails ignited on the Boulevard Saint-Michel, Lefebvre assumed the role of a radical honest broker, trying to galvanize the "old" Left—his generation, who tended to rally around class, party, and trade unions—with an emergent "New Left," a younger crew of militants, less steeped in theory, who organized around anti-imperialism and identity themes and who spoke the language of culture and everyday life. The parallels with post-Seattle agitation are striking. The Lefebvrian desire to conjoin young and old progressives around a concerted anticapitalist struggle remains as pressing and as instructive as ever; his theories about space equally resonate within analyses of globalization, just as his notion of the "urban revolution" and "right to the city" endure as visionary democratic ideals. Lefebvre warned us long ago that the ruling class will always try to suppress and co-opt contestation, will always try to convert romantic possibility into realistic actuality. He knew that in desiring the impossible, in reaching for the stars, we might at least one day stand upright.

His was a praxis that borrowed more from Rosa Luxemburg than Vladimir Lenin, whiffed of Norman O. Brown rather than stank of Leonid Brezhnev. In the 1970s, somebody asked Lefebvre if, in fact, he was really an anarchist. "No," he replied. "I'm a Marxist, of course, so that one day we can all become anarchists!"[12] His Marxism was unashamedly Rabelaisian, nurtured in the fields as well as in the factories, festive and rambunctious, prioritizing "lived moments," irruptive acts of contestation: building occupations and street demos, free expressionist art and theater, flying pickets, rent strikes, and a general strike. Here the action might be serious—sometimes deadly serious—or playful. Lefebvre dug the idea of politics as festival. Rural festal traditions, he said in *Critique of Everyday Life* (1947), "tighten social links at the same time as they give free rein to all desires which have been pent up by collective discipline and necessities of work." Festivals represent "Dionysiac life … differing from everyday life only in the explosion of forces which had been slowly accumulating in and via everyday life itself."[13]

* * *

A few summers ago, I decided to check out Lefebvre's Dionysian roots for myself. One warm July evening I arrived at the village where he'd grown up and vacationed during college recesses. I wasn't sure exactly what I was looking for in Navarrenx, or what I'd find, but I knew somehow the pilgrimage would help me better understand the man himself, and his milieu. Sure enough, I realized immediately I'd discovered the rustic ribald body to the Parisian professor's cool analytical head. A marvel of Middle Age town planning aside the River Oloron, in the foothills of the Pyrénées-Atlantique, the *bastide* of Navarrenx remains charming, sleepy, and just about vital five centuries on. Imposing ramparts with two ancient town gates—Porte Saint-Germain and Porte Saint-Antoine—encircle its grid pattern of higgledy-piggledy streets that are today lined with a few *boucheries* and *boulangeries,* the odd melancholy café, and several pizzerias. Those walls hark back to 1537, when the King of Navarre refortified the fourteenth-century originals. Thirty-odd years later, Navarrenx, whose name has Basque origins, withstood a three-month siege defending the honor of Jeanne d'Albret, sovereign of Béarn and mother of King Henri IV. Two centuries on, in 1774, the town underwent extensive renovation and replanning; many structures, including *chez* Lefebvre at rue Saint-Germain, hail from this period.

At the nearby Place des Casernes, an almost deserted square shadowed by the Porte Saint-Antoine, the gateway to Spain, modern-day travelers can find no-frills room and board at Navarrenx's sole inn, the Hôtel du Commerce. My first, and only, evening at the Hôtel du Commerce seems comical in retrospect. I'd decided to take a twilight stroll along Navarrenx's ramparts, imbibe its atmosphere in the balmy air. When darkness fell, I returned to find my room infested with mosquitoes; the South West's damp, mild climate is a veritable breeding ground for these pests, and I'd dumbly left the light on and shutters open. Too late for room service, I chose the fastest remedy: to splatter every single one with a rolled-up newspaper. Next morning, in broad daylight, I realized the mess I'd made to the walls and ceiling, much to the chagrin of Monsieur *le propriétaire,* who wasn't amused. We exchanged words; I placated, apologized, promised to clean everything up, which I hastily did. Yet the portly *patron* wasn't impressed and urged me to pay up and clear out, sooner rather than later. Thus, like a renegade pilgrim of Saint-Jacques de Compostelle, I was banished from Navarrenx, kicked out on my debut visit.

The banishment had been a strange blessing. Forced to flee, I discovered the Basque town of Mauléon, twenty minutes down the road, and the wonderful Bidegain hotel, which serves the lovely rich, deep-bodied *Irouléguy* wine Lefebvre tippled.[14] As the signature red shutters and Basque red, green, and white flags became more prominent, I saw and felt the proximity of Navarrenx to Basque county; I began to grasp up close how its culture and tradition affected Lefebvre's own spirit and personality. His "fanatically religious" mother was of Basque stock. " 'You speak against religion,' she and her sisters scorned me. 'You will go to hell.' "[15] Lefebvre recognized the contradictions traversing Basque culture because those same contradictions traversed him: the Basques "have a very profound sense of sin; and yet, they love to live, love to eat and drink. This contradiction is irresolvable, because it's a fact I've often stated: the sense of sin excites pleasure. The greater the sin, the greater also the pleasure."[16] His libertine roots lay on his father's side, a Breton free spirit who loved to gamble and usually lost. "My Breton father bequeathed me a robust and stocky body [*trapu*] … [he was] of light, easy mood, Voltairean and anticlerical. … I believe that from birth that I resembled him."[17] He inherited his mother's facial features: "a long, almost Iberian face." "The head of Don Quixote and the body of Sancho Panza," one lady friend described Lefebvre; she knew him well. "The formula," he said, "hadn't displeased me."[18]

Inside Lefebvre's body and mind lay a complex dialectic of particularity and generality, of Eros and Logos, of place and space; he was a Catholic country boy who had roamed Pyrenean meadows, a sophisticated Parisian philosopher who'd discoursed on Nietzsche and the death of God. He was rooted in the South West yet in love with Paris, tormented by a Marxist penchant for global consciousness. This triple allegiance tempered hometown excesses, made him a futuristic man with a foot in the past, someone who distanced himself from regional separatism. "Today," Lefebvre warned, "certain [Basque] pose the question of a rupture with France. I see, in regionalism, the risk of being imprisoned in particularity. I can't follow them that far. … One is never, in effect, only Basque … but French, European, inhabitant of planet earth, and a good deal else to boot. The modern identity can only be contradictory and assumed as such. It also implies global consciousness."[19] The incarnation of a man of tradition and a Joycean everyman is suggestive in an age that frantically invokes an essential purity of identity or else wants to homogenize everything in a nihilistic market rage.

"I know nothing better in the world than this region," said Lefebvre in *Pyrénées,* his alternative tour guide of well-trodden paths. "I know its strengths and weaknesses, its qualities and faults, its horizons and limits. … I have savored the earth in my lips, in the breeze I've smelt its odors and perfumes. The mud and stones and grassy knolls, the peaks and troughs of its mountains, I've felt them all underfoot."[20] Written as part of the "Atlas des Voyages" series, this brilliantly poetic travelogue, both geographically materialist and romantically lyrical, mixes photos of dramatic Pyrenean landscapes and ruddy-faced peasants with citations from Hölderlin and Elisée Reclus. Meanwhile, we can glimpse the "crucified sun," those crucifixes so ubiquitous in the South West's landscape—giant, austere crosses framed against a bare circle symbolizing the sun. They'd put the fear of God in anyone. Lefebvre equated such religious iconography with bodily repression and ideological dogmatism; it's an imagery and mentality, in whatever guise, he'd spend a lifetime shrugging off and battling against. "I understand the Pyrenean region better than anyone," Lefebvre claimed, better than its inhabitants, "precisely because I quit it for elsewhere. … No, not just for Paris, but elsewhere in my consciousness and thought, elsewhere in the world; elsewhere in 'globality,' in Marxism, in philosophy, in the diverse human sciences."[21]

* * *

Last fall, I went to seek out another little piece of Lefebvre's world, in an unlikely place: the rare book archives of Columbia University's Butler Library in New York. There, you can find the one-hundred-odd letters Lefebvre sent his longtime friend and collaborator Norbert Guterman. Guterman, a Jew, left France in 1933 and settled permanently in New York. With Lefebvre and writer–communists Paul Nizan (gunned down in Dunkirk in 1940, at age thirty-five) and Georges Politzer (who could swear in German at his Nazi torturers), Guterman collaborated on a series of short-lived philosophy journals. For more than forty years, until Guterman's death in 1984, he and Lefebvre corresponded. In 1935 they busied themselves on a book, trying to explain why, despite being counter to its collective interests, the German working class ran with Hitler. Lefebvre and Guterman appealed for a Popular Front that could reconcile fractional differences and catalyze a *gauchiste* revolution.

Alas, their book, published the following year, was denounced in "official" communist circles and dismissed as Hegelian and revisionist. Yet the thesis survived the party and the plague, and its intriguing title is apt for explaining the *zeitgeist* of contemporary America seventy years down the line: *La Conscience Mystifiée—mystified consciousness,* a consciousness not only usurped by the fetishism of the market but *alienated* from itself by "absolute truths" of nationhood, patriotism, God, and the president. Lefebvre's letters from this period are shadowed by a pessimism of impending doom that has a familiar ring about it: "a funk prevents the people from thinking and living," he wrote on October 19, 1935; "The moment of catastrophe approaches," recalled another communiqué (January 1936). "I will not make a will," Lefebvre confessed, on the brink (August 28, 1939). "What I would have been able to bequeath isn't yet born. … I don't think of posterity in writing to you, but of our work, our fraternity, our true friendship." The "Guterman Collection" is a moving testimony of an enduring friendship that survived a century of war and peace, love and hate, displacement and disruption.

Yellowing letters, written on tissue-paper parchment, on regional Communist Party notepaper ("*La Voix du Midi*"), on postcards mailed from Algeria, Greece, Italy, Brazil, and Spain—all bear Lefebvre's typical cursive: free flowing and fast paced, spread frantically and unevenly across the page. His pace mimics both political mood and personal circumstance. They confide in each other. "Mon cher vieux Norbert," many of Lefebvre's letters begin, affectionately. "I would love to know what you're doing, and how you live in America." Lefebvre bemoans his dire family situation during the Occupation, his penury after the peace, his struggles to find a steady teaching job, his latest love: Evelyne, Nicole, Catherine, whose own letters crop up in the archive. "I spend my time," explains Evelyne to Norbert, "typing what Henri has feverishly written to earn us a few sous."

In another letter, dated October 18, 1977, Lefebvre said, "I almost forgot to tell you that Catherine [Regulier] and me are making a book together: a series of philosophical and political dialogues between a very young woman and a monsieur who has no more than a youthfulness of heart." The eventual text, *La révolution n'est pas ce qu'elle était* (1978)—"the revolution isn't what it used to be"—expressed Lefebvre's open-ended, inventive Marxist spirit, which continually updated itself as society updated itself. It's a spirit we can still tap. Indeed, as the sclerosis of our body politic hardens to the point of apoplexy, we need, perhaps more desperately than ever, not only a new Popular Front but also a certain monsieur's "youthfulness of heart."

* * *

Henri Lefebvre: A Critical Introduction tries to resuscitate the sweeping style and youthful spirit of Henri Lefebvre, metaphilosopher, *bon vivant,* utopian. In what follows, I consider the man, his work, and his ideals and bring each to bear on a culture that seems intent on throwing itself down a deep and dark abyss. His heterodox Marxist rigor, his optimism of the intellect as well as the will, his frank concern for profane human happiness all seem *especially* inspiring in an era when crony philistinism has supposedly rendered such a "meta-style" old hat. Lefebvre was a thinker whose life and thought progressed in a kind of episodic and peripatetic unison. His oeuvre became not just something written down on paper but a reality actually *lived.*

The present offering explores, more modestly, what Lefebvre's own *La Somme et le Reste* explored: the sum and the remainder, recounting a tale of what Lefebvre achieved while pioneering the way for the "rest," for the still-to-be-accomplished, the still-to-be-lived aspect of that legacy. In the chapters to follow, I probe key concepts: Everyday Life (chapter 1), Moments (chapter 2), Spontaneity (chapter 3), Urbanity and the Urban Revolution (chapters 4 and 5, respectively), Space (chapter 6), Globalization and the State (chapter 7), Mystified Consciousness (chapter 8), and the Total Man (afterword), in the light (and darkness) of the present conjuncture. As they shift thematically, each chapter will *periodize* a specific facet of Lefebvre's life and thought at the same time as it tries to stress how these particular facets live on today, as an enduring interrelated whole. Specific chapters can be read alone, as discrete themes, but I'd like to stress their interweaving and overlapping nature—their "*s'entrelacer,*" as Lefebvre might have said.

In the Anglophone world, geographers, urbanists, and cultural theorists have appropriated Lefebvre as their own during the past decade or so. There, *The Production of Space,* perhaps Lefebvre's best-known book, is one of his least-known texts in the Francophone world, who generally acknowledge Lefebvre as a Marxist philosopher cum rural–urban sociologist; in this camp Lefebvre reigns as a prophet of alienation and Marxist humanism, a thinker who brought an accessible Marx to a whole generation of French scholars. (*Le Marxisme* [1948], appearing in the immensely popular "*Que sais-je?*" series—"What do I know?"—remains far and away his best-selling book.) And yet, only two of Lefebvre's many Marxist monographs (*The Sociology of Marx* and *Dialectical Materialism*) have made it into English; his greatest work, *La Somme et le Reste,* is still also untranslated; ditto his pathbreaking (and relevant) exploration on "mystified consciousness"; ditto his work on the state, Nietzsche, and Rabelais; ditto his critique of technocratic culture and treatises on aesthetics and representation.

Anglo-American studies that see Lefebvre as a preeminent spatial thinker and urbanist—themes he only began to pick up as a sexagenarian—often overlook the fact that he was first of all a Marxist. Texts that discuss his concept of everyday life tend to make short shrift of his dialectical method and utopian "total man," thereby severing parts of an oeuvre that coexist in dynamic unity. To this degree, a thinker who detested compartmentalization has been hacked apart and compartmentalized within assorted academic disciplines. For that reason, I want to keep together Lefebvre's Francophone and Anglophone strands, highlighting the interrelatedness of his scholarship, its polemical edge and playful twists, its everyday aspects and other-worldly yearnings, its realism and its surrealism. I want to travel, as Lefebvre traveled, through time and over space, engaging with his times as well as our own. En route, I hope this little book reaches the reader as a *critical introduction,* one that emphasizes the *Oxford English Dictionary*'s other notion of "critical": it will introduce Henri Lefebvre *critically,* at a "*critical* moment," when it is "decisive and crucial" to do so, when his ideas and politics are "of *critical* importance."

Note: Whenever and wherever possible, the author makes use in this book of existing English translations of Lefebvre. Elsewhere, all citations from the original French are translations made by the author.

1

Everyday Life

> One finds all one wants in the *Grand Magasins* of everyday adventure, which never close, even on Sundays and holidays.
>
> —**Pierre Mac Orlan,** *Chroniques de la fin d'un monde*

It's astonishing to think that Henri Lefebvre began Volume 1 of *Critique of Everyday Life* with the founding of the United Nations and finished it with Volume 3, in 1981, during the first term of Ronald Reagan. In between, in 1961, just as mass consumerism really took off, he penned Volume 2. (He also wrote, as some of his students barricaded Paris's boulevards, *Everyday Life in the Modern World*.) It was quite a stretch, quite a project: beginning in the age of peace and consensus, continuing through a cold war and a counterculture, and sealing it amidst a neocon backlash. His opening salvo in 1947 was that of a man of the countryside, even though he found himself back pacing Paris's streets, working for the Centre National de Recherche Scientifique (CNRS), after a period teaching in the provinces.[1] At CNRS, Lefebvre focused on the peasant question, conducting research on agricultural reform in France, Italy, and Eastern Europe and on "primitive accumulation" of capital, as well as on the rural rent issues that Marx left dangling in Volume 3 of *Capital*. Lefebvre always felt that the peasantry figured prominently in socialist history; Mao's 1949 revolution in China offered dramatic confirmation. (The French Communist Party, though, was less impressed with poor Lefebvre's peasant labors. Rural rent, they scoffed, was a Ricardian problematic not a Marxist one!)

An even more amazing aspect of Lefebvre's notion of everyday life, one overlooked by many commentators, is that it germinated when everybody's daily life, Lefebvre's included, was about to be blown to smithereens. Therein lies its most fundamental message: everyday life is so precious because it is so fragile; we must live it to the full, inhabit it as fully sensual beings, as total men and women, commandeering our own very finite destiny, before it's too late. The life and death everyday drama for Lefebvre really began in December 1940, when he quit his teaching post as "a little *prof de philo* in a little provincial *collège*" (high school) at Montargis, one hundred kilometers south of Paris, and accepted another at Saint-Étienne, further south in the Loire.[2] Married with four kids, Lefebvre's already fraught personal situation soon worsened when the pro-Nazi Vichy government began purging public offices, schools, and colleges of Jews, Freemasons, and Communist Party members. Too old to be drafted, without job or means, the almost fortysomething philosopher fled to Aix-en-Provence, where he joined the Resistance Movement and lived in a tiny house a few kilometers out of town. In winter, it was freezing cold. For fuel he burned wood that created more fumes than warmth, bringing on a bout of bronchitis; the ailment periodically recurred throughout his life.[3]

At Aix's *Café Mirabeau,* Lefebvre met other *maquisands,* organized clandestine conspiracies and sabotage, and befriended railway men who helped him derail enemy trains and sniff out collaborators.[4] "We worked to give an ideology to the Resistance," Lefebvre remembers. "Vichy held up the flag of Revolution and Empire and said to the Germans that they'd guard the colonies for Hitler. … In Vichy, there'd been those who sincerely believed in preserving the independence of a part of France, controlled between Germany and a zone to the south. … The Resistance explained that this independence was a fiction."[5] Lefebvre also descended regularly on Marseille, the real hotbed of struggle, and frequented the café *Au Brûleur de Loup,* where militant wolves, free-spirit wanderers, on-the-run refugees, and those seeking departure for America all found warm sanctuary. Surrealist André Breton hung out there before sailing to New York; ditto Victor Serge, the Russian anarchist and veteran revolutionary, who later eloped to Martinique. In Marseilles, Lefebvre befriended Simone Weil, the devout philosopher–martyr; he was pained as he watched her battle for interns in nearby camps while starving herself to death. (Weil eventually died of tuberculosis in a Kent sanatorium in England in 1943.)

In *Memoirs of a Revolutionary,* Victor Serge lets us feel the spirit and guts of those times, of the Frenchmen, whether intellectuals or workers, who had no intention of emigrating. "Various militants tell me," Serge said, "quite simply, 'Our place is here,' and they were right."[6] But André Breton opted to leave just as Lefebvre risked life and limb to stay. To visit his parents back in Navarrenx he made daring, stealth night raids. They were terrified for their son, and for themselves; somebody might see him, somebody might inform on him, and on them. He went underground, and then, at the beginning of 1943, Lefebvre hid himself away in an isolated Pyrenean peasant community in the valley of Campan, near Tarbes. He laid low with locals, and with local *maquisands,* until the Liberation. He got to know mountain shepherds on the slopes, studied them, learned their rituals and folklore and *façon de vivre,* and even spotted a sort of primitive communism in their daily life. He didn't know it then, but he'd already embarked on everyday life research, pregnant in his doctorate on peasant sociology, *Les Communautés Paysannes Pyrénéennes* (eventually defended in Paris in June 1954).[7]

Methodologically, Lefebvre deployed a sort of "participant observation," which, coupled with long sessions in the archives of Campan's Town Hall, led him to discover a passion for historical excavation he never knew he had. Jean-Paul Sartre, for one, appreciated the virtues of Lefebvre's rural "regressive–progressive" methodology—a methodology informing his work on urbanism and space decades later. "In order to study complexity and reciprocity of interrelations—without getting lost in it—Lefebvre," Sartre noted, "proposes 'a very simple method employing auxiliary techniques and comprising several phases: (a) *Descriptive.* Observation but with a scrutiny guided by experience and a general theory. … (b) *Analytico-Regressive.* Analysis of reality. Attempt to *date* it precisely. … (c) *Historical-Genetic.* Attempt to rediscover the present, but elucidated, understood, explained.' " "We have nothing to add to this passage," Sartre added, "so clear and so rich, except that we believe that this method, with its phase of phenomenological description and its double movement of regression followed by progress, is valid—with the modifications which its objects may impose upon it—*in all the domains of anthropology.*"[8]

As Lefebvre documented the plight of the rural peasant and the agrarian question under socialism, his "critique of everyday life" took shape. After 1947, this became both a methodology and a political credo: an insistence that dialectical method and the

Marxist dialectician confront the everyday, that they begin and end analysis in the quotidian. For Lefebvre, everyday life became a bit like quantum theory: by going small, by delving into the atomic structure of life as it is really lived, you can understand the whole structure of the human universe. A politics that isn't everyday, Lefebvre says, is a politics without a constituency. Therein lay the problems of party Marxism, with its preoccupation with building an abstract economy rather than reinventing a real life. On the other hand, an everyday life without historical memory, without any broader notion of its dialectical *presentness,* is forever prey to mystification. "When the new man has finally killed magic off," Lefebvre says in Volume 1 of *Critique of Everyday Life,* with trademark rhetorical flush, "and buried the rotting corpses of the old 'myths'—when he is on the way towards a coherent unity and consciousness, when he can begin the conquest of his own life, rediscovering or creating *greatness in everyday life*—and when he can begin knowing it and speaking it, then and only then will we be in a new era."[9]

* * *

Much in *Critique of Everyday Life* seemed like light relief, like Lefebvre's romp through cherished books and sunny, open meadows. He seems deliberately to want to put those war years aside, out of sight and out of mind. His debut volume is discursive, free flowing, and formless—a welter of ideas and muses, allusions and alliterations, spiced up with playful doses of polemicism. At times, we have to work hard to keep up. He gives us a recapitulation of "some well-trodden ground," reconsidering questions about alienation and surrealism: André Breton's clarion call, Lefebvre jokes, is "Snobs of the World Unite!" Once again, he tussles with the party, defending humanism and "Marxism as Critical Knowledge of Everyday Life." A lot unfolds like a stream of consciousness, as Lefebvre breezes through the "French countryside on a Sunday afternoon," demystifying the "strange power" of a village church—a church that could exist anywhere today: "O Church, O Church, when I finally managed to escape from your control I asked myself where your power came from. Now I can see through your sordid secrets. ... Now I can see the fearful depths, the fearful reality of human alienation! O holy Church, for centuries you have tapped and accumulated every illusion, every fiction, every vain hope, every frustration."[10]

Elsewhere, Lefebvre juggles with this concept he labels "everyday life," typically weary of laying it down solid. Literature and art, he says, as opposed to politics and philosophy, have better grappled with understanding the everyday.[11] Brecht's "epic drama" gives us a theater of the everyday, where all the action is stripped of ostentation and where all truth, as Brecht liked to say, citing Hegel, "is concrete." "Epic theater," Lefebvre quotes Brecht preaching, "wants to establish its basic model at the street corner." Brecht has his great hero of knowledge, Galileo, begin by a process of "de-heroization": "GALILEO (*washing the upper part of his body, puffing, and good-humored*): Put the milk on the table."[12]

The films of Charlie Chaplin, meanwhile, whose image of the tramp strike as both "Other" and universal in "modern times," reveals bundles about everyday alienation, and, just like life itself, its drama is a slapstick that makes us laugh and cry, sometimes at the same time. (In the 1950s, Chaplin and Brecht both felt the heat from Senator Joseph McCarthy's "red" witch hunts. Their power to disgruntle and critically inform was thereby acknowledged.) Chaplin, according to Lefebvre, "captures our own attitude towards these trivial things, and before our very eyes."

> He comes as a stranger into the familiar world, he wends his way through it, not without wreaking joyful damage. Suddenly he disorientates us, but only to show us what we are when faced with objects; and these objects become suddenly alien, the

> familiar is no longer familiar (as for example when we arrive in a hotel room, or a furnished house, and trip over furniture, and struggle to get the coffee grinder to work). But via this deviation through disorientation and strangeness, Chaplin reconciles us on a higher level with ourselves, with things and with the humanized world of things.[13]

The other brilliant spokesperson of the everyday is, of course, James Joyce. His masterpiece *Ulysses,* Lefebvre notes, "demonstrates that a great novel can be boring. And 'profoundly boring.' Joyce nevertheless understood one thing: that the report of a day in the life of an ordinary man had to be predominantly in the epic mode."[14] The bond between Leopold Bloom, one ordinary man during a single, ordinary day in Dublin, and the heroic epic journey of Odysseus is precisely the bond that exists between Lefebvre's ordinary man and his "total man," between the present and the possible. The former is pregnant with the latter, already exists in the former, in latent embryonic state, waiting for Immaculate Conception, for the great, epochal imaginative leap. Thus, while Lefebvre's utopian vision of the total man seems way out, and grabs us an idealist mixture of hope and wishful thinking, his model is really anybody anywhere, any old Leopold or Molly Bloom or Stephen Dedalus. What appears to be stunningly abstract is, in reality, mundanely concrete: the ordinary is epic just as the epic is ordinary. In *Ulysses,* "Blephen" and "Stoom" find a unity of metaphysical disunity, just as the ordinary man and total man can find their unity of metaphysical disunity; the poet–artist son and the practical-man-of-the-world father conjoin. Two world-historical temperaments—the scientific and the artistic—become one and soon wander empty darkened streets, wending their way back home to where Molly sleeps in Ithaca, at 7 Eccles Street.

In a stunning literary, psychological, and—perhaps—revolutionary denouement, *Ulysses* ends with Molly's tremendous stream of unpunctuated consciousness; visions and opinions, fragments and perceptions, judgments and recollections gush forth in one of modern literature's greatest set pieces. He "kissed me under the Moorish wall and I thought well as well him as another … would I yes to say yes my mountain flower and first put my arms around him yes and drew him down to me so he could feel my breasts all perfume yes and his heart was going like mad and yes I said yes I will Yes."[15] The *Ulysses* that says Yes to life is an "eternal affirmation of the spirit of man," a great gust of generosity that is indeed the spirit of Lefebvre's total man. Yet Lefebvre knew it bespoke a more commonplace theme: everyday passion. These, both he and Joyce knew, match the dramatic successes and failures of Greek heroes. Life at its most mundane level is as epic and spiritual as any official history or religion. History, as Stephen reminds his boss Mr. Deasy, the bigoted, protofascist headmaster, is really "a shout in the street." Lefebvre, the Marxist everyman, would doubtless concur: total men and women are found on a block near you.

Lefebvre's sensitivity to everyday life also smacks as a French thing. The daily round is deeply ingrained in French culture where rhythms and rituals punctuate and animate places and people everywhere: the early morning stroll to pick up the bread;[16] the first cup of coffee; a meal at lunchtime for which everything closes down and families still commingle; a sip of wine and a piece of cheese; the chime of a church bell on the hour; the familiar bark of a neighborhood dog; a *Café du Commerce* almost anywhere, frequented by a loyal clientele who appear at the same hour each day—simple, ostensibly trivial occurrences that assume epic proportions. As novelist Pierre Mac Orlan once put it in a perceptive memoir called *Villes,* and as Lefebvre equally comprehended, "It is the finest quality of the French that they can render agreeable a block of houses, a few farms, two or three lamplights, and a sad café where you die of boredom playing dominoes. It isn't so much that, on this vast earth, the French are nicer than anybody else, but more that they know how to bring a bit of pleasantry to their little existence."[17]

This is the positive aspect of daily life: it was familiar, it was the realm of home and leisure, the arena of safety and security, of friends and families, of holidays and little treats—the side of life that included work but was somehow separated from work, set aside from work, liberated from it. Alienation that pervaded at the workplace hadn't yet penetrated the everyday, the nonspecialized activities that lurked outside the factory gate and the office cubicle, beyond the school staffroom or store checkout. Or had it? It's now that Lefebvre asserts his Marxist credentials; it's here where the negative side of daily life emerges. For more and more, he said in 1947, prophetically, everyday life was being *colonized*. Colonized by what, exactly? Colonized by the commodity, by a "modern" postwar capitalism that had continued to exploit and alienate at the workplace but had now begun to seize the opportunity of entering life in general, into nonworking life, into reproduction and leisure, free time and vacation time. Indeed, it was a system ready to flourish through consumerism, seduce by means of new media and advertising, intervene through state bureaucracies and planning agencies, ambush people around every corner with billboards and bulletins. And it would boom out in the millions of households that possessed TVs and radios.

* * *

In 1958, Lefebvre drafted a long foreword to his 1947 original text, evaluating the state of the game ten years on. The pincers of a cold war and a capitalist consumerist war squeezed tighter and tighter. On the one side, state socialism bureaucratized daily life, planned and impoverished it, converted it into a giant factory intent on productive growth; on the other side, state capitalism ripped off everyday life and sponsored monopoly enterprises to mass produce commodities and lifestyles, dreams and desires. One system transformed the realm of freedom into the drudge of necessity; the other turned a repetitive necessity into a supposed freedom. Yet *Critique of Everyday Life* aimed to get inside both systems, expose their pitfalls, journey beyond them. "It is ludicrous to define socialism solely by the development of the productive forces," Lefebvre writes. "Economic statistics cannot answer the question: 'What is socialism?' Men do not fight and die for tons of steel, or for tanks and atomic bombs. They aspire to be happy, not to produce."[18] They aspire to be free, not to work—or else to work less. Thus, Arthur Rimbaud's provocative plea of the "right to be lazy" is a right socialism needed to reconcile.[19] Changing life, inventing a new society, can be defined only, Lefebvre says, "*concretely* on the level of everyday life, as a system of changes in what can be called lived experience."[20] But here, too, lived experience was changing in advanced capitalist countries; it was under fire from forces intent on business and market expansion, producing fast cars and smart suburban houses, consumer durables and convenience food, processed lives and privatized paradises.

As such, everyday life possessed a dialectical and ambiguous character. On the one hand, it's the realm increasingly colonized by the commodity, and hence shrouded in all kinds of mystification, fetishism, and alienation. "The most extraordinary things are also the most everyday," Lefebvre quips, reiterating Marx's comments on the "fetishism of commodities" from *Capital 1;* "the strangest things are often the most trivial."[21] On the other hand, paradoxically, everyday life is a primal arena for meaningful social change—the only arena—"an inevitable starting point for the realization of the possible."[22] Or, more flamboyantly, "everyday life is the supreme court where wisdom, knowledge and power are brought to judgment."[23] Nobody can get beyond everyday life, which literally internalizes global capitalism, just as global capitalism is nothing without many everyday lives, lives of real people in real time and space. Lefebvre is adamant that a lot of Marxists held a blinkered notion of class struggle, a largely abstract and idealized version that neglects, he reckons, not only "recent modifications of capitalism" but also the "socialization of production."[24] Put differently, Marxists had let the world pass them by; rather than confront the mundane realities of modern everyday life, they'd turned their backs away from them.

In the "Economic and Philosophical Manuscripts," Marx said a worker "does not confirm himself in his work, but denies himself, feels miserable and not happy, does not develop free mental and physical energy, but mortifies his flesh and ruins his mind. Hence the worker feels himself only when he is not working; when he is working he does not feel himself. He is at home when he is not working, and not at home when he is working."[25] Lefebvre suggests workers no longer feel at home even when they're not working; they're no longer themselves at home, given that work and home, production and reproduction—the totality of daily life—have been subsumed, colonized, and invaded by exchange value. For leisure, workers give back their hard-earned cash as consumers, as mere bearers of money; private life, meanwhile, becomes the domain where they're lured to spend, the domain of the ad, of fashion, of movie and pop stars and glamorous soap operas, of dreaming for what you already know is available, at a cost. In an ever-expanding postwar capitalism, all boundaries between economic, political, and private life are duly dissolving. All consumable time and space is raw material for new products, for new commodities. Marx's "estranged labor" now generalized into an "estranged life."

Everyday life mimics in the social realm what Marx spotted in the economic realm: the notion of social man being conditioned by a system that produced through private labor; that society was founded on a privatized basis; and that never could capitalism square the circle, humanize the social by means of the private. Once, everyday life was textured around the private and the social realms, a private consciousness and a social consciousness. At home, family life could be private without being *deprived;* when people flocked onto the street and into cafés, went to a town meeting or a community event, their everyday day was participatory, socialized. The two realms coexisted within a unity where money relations were a conditioner rather than a determinant. Yet with the dissolution between work and leisure, and with the expansion of exchange value into the "totality of daily life," this fragile unity was severed, and both flanks suffered. Now, either at home or at work, in the private or the public realm, commodities and money, gadgets and multimedia, reign supreme.

Now, lived experience is both colonized economically and usurped ideologically: it rocked to the beat of a *conscience privée,* wallowed in mystification, reveled in its own *deprived consciousness.* Lefebvre's old thesis from *La Conscience Mystifiée* (1936) returns to expose not modern fascism but modern capitalism—in its everyday, trivial guise. The stakes are intensified once machines and technological knowledge burst on to the scene. But what's going on is more than the application of technology at the workplace. As Marx pointed out, technology "reveals the active relation of man to nature, the direct process of the production of his life, and thereby it also lays bare the process of the production of the social relations of his life, and of the mental conceptions that flow from those relations."[26]

Changes in the means of production transform our mode of life and, in turn, transform the ideas we have about our world and ourselves. Think about how the human brain invented the steam engine, Fordist mass production, space travel, biotechnology, e-mail, and the Internet. But think about how these have equally invented us, successively shaped the way we look at ourselves. For Marx and Lefebvre alike, these instruments of man only betoken man the instrument. In effect, machines lessen the burden. In reality, they become an "alien power," more frantically setting in motion labor power, transforming people into mere appendages of mechanical devices, crippling true subjectivity, and ushering in the "real subsumption" of everyday life under the domain of capital. The workweek continues to grow longer and longer in the technologically most advanced nation in the world, the United States, despite—or because of—time-saving ingenuity.

Who'd be surprised, given that cellular phones, e-mail, laptops, and various handheld electronic devices permit many people to work while they're traveling to work and to work at home, at their leisure. For the lucky ones who can labor at home or on the beach, in hotels or at airports—as the unlucky ones toil at multiple jobs to keep daily life afloat—it's hard to know whether these changes represent absolute worker empowerment or total enslavement. Is this high-tech, liberated labor force a new industrial aristocracy, or has capitalism, as Marx pointed out in the *Manifesto,* "stripped of its halo every occupation hitherto honored and looked up to with reverent awe? It has converted the physician, the lawyer, the priest, the poet, the man of science, into its paid-laborers."[27] Either way, the *gadget* has permeated new millennium daily life, filled in the unproductive pores of the working day, created human personalities permanently online, addictively tuned in, programmed to perform, and terrified to log off. A tiny Nokia object, stuck in somebody's ear, now represents a curious alien power, a heady narcotic that underwrites the rhythms and texturing of people's everyday life. Every civic space, every street or café, assumes the quality of a surrogate living room—or an open-planned office, a postmodern relay system.

* * *

For Lefebvre, the contradictions of everyday life inevitably find their solutions in everyday life. How could they otherwise? Grappling for answers, he journeys a little closer to home, looks over his shoulder, and remembers his roots. Since childhood he'd known a tradition that is the veritable nemesis of insurgent forms of modern alienation: *the rural festival.* The drama usually ended in rowdy scuffles and raving orgies; festival days were rough and tumble and full of vitality, and Lefebvre loved them and romanticized them in adulthood. (Pieter Brueghel's painting *Battle of Carnival and Lent* magnificently portrays this raucous medieval lifeworld.) Festivals seeped into Lefebvre's Marxist conscience, activated involuntary memory, and aroused primordial visions of infant paradise, tasting a little like a Proustian *madeleine* dipped in tea; the sensation recreated the past, only to unlock the Pandora's box of the future. Lefebvre's philosophical homesickness locates itself in the future, and the past becomes a platform for pushing forward, partying onward, toward a higher plane of critical thinking and practice. He saw in festivals paradigms of an authentic everyday life, a realm where the shackles of enslavement had been loosened.

Indeed, festivals "tightened social links," he says, "and at the same time gave rein to all the desires which had been pent up by collective discipline and the necessities of everyday work. In celebrating, each member of the community went beyond themselves, so to speak, and in one fell swoop drew all that was energetic, pleasurable and possible from nature, food, social life and their own body and mind."[28] Lefebvre invokes the festival during the 1940s and 1950s as a jarring antithesis of bureaucratic domination and systematized ordering. Like Faust, he fraternizes with the demonic and gives himself over to Dionysius, to excess and *un*productivity, to Eros rather than Logos, to desire rather than depression. Festivals were like everyday life, only more intense, more graphic, more raw. During festivals, people dropped their veils and stopped performing, ignored authority and let rip. They broke out of everyday life by affirming what was already dormant in everyday life—and dormant in themselves. Festivals "differed from everyday life," sometimes "contrasted violently with everyday life," but, and this is a big *but* for Lefebvre, "*they were not separate from it*."[29] On the contrary, they "differed from everyday life only in the explosion of forces which had been slowly accumulated in and via everyday life itself."

Lefebvre's penchant for festivals was catalyzed by that maestro *fêtard,* François Rabelais, the sixteenth-century poet–sage, who, in his sprawling, magical–realist masterpiece *Gargantua and Pantagruel* (1532–56), created a whole literary and philosophical edifice based on wine and eating, carnivals and laughter. Rabelais's mockery of Middle Age authority, Lefebvre maintained in his 1955 study *Rabelais,* can help us mock our own authority and our own contemporary seriousness, and restore a new sense of democracy and lighter meaning to everyday life. Here play and laughter become revitalized seriousness, no joking matters, not sidetracks and diversions to making money and accumulating commodities. In the bawdy and biting *Gargantua and Pantagruel,* with its great feasts of food and drink, rambunctious reveling and coarse humor, Rabelais denounced all forms of hypocrisy. "Readers, friends," he warned his audience—old and modern alike—"if you turn these pages / Put your prejudice aside, / For, really, there's nothing here that's contagious. / Nothing sick, or bad—or contagious. / Not that I sit here glowing with pride / For my book: all you'll find is laughter: That's all the glory my heart is after, / Seeing how sorrow eats you, defeats you. / I'd rather write about laughing than crying, / For laughter makes men human, and courageous." "BE HAPPY!" Rabelais urged.[30]

Lefebvre presents Rabelais as a visionary realist who has a foot in the past as well as an inkling of the future—of the contradictory birth bangs of modern capitalism, the new mode of production invading his old world. In an odd way, Rabelais also propels us into a postcapitalist world, because, Lefebvre argues, he revealed a "vision of the possible human, half-dream, half-fantasy … an *idea of a human being*."[31] Lefebvre's *Rabelais* finds *son semblable, son frère* within the leaves of another Rabelaisian prophet, the Russian critic Mikhail Bakhtin, whose study *Rabelais and His World* elevated Rabelais to the summit of the history of laughter.[32] Rabelaisian laughter was intimately tied to freedom, Bakhtin similarly argued, especially to the courage needed to establish and safeguard it. Written in the 1930s, during the long nights of Stalin's purges, *Rabelais and His World* endorsed the spirit of freedom when it was increasingly being suppressed. Bakhtin's text wasn't translated into French until 1970 and so was unread by Lefebvre in 1955; it came to English audiences in that big party year of 1968. Bakhtin's closest contemporary would have been a book and a theorist Lefebvre did actually know: *Homo Ludens* (1938)—"Man the Player"—by the Dutch medieval historian Johan Huizinga, who emphasized the play element in Western culture just as Hitler got deadly serious across Europe.

Like Bakhtin and Huizinga, Lefebvre adores Rabelais's laugher, but his laughing Rabelais guffawed as a probing critic. Lefebvre's Rabelais chronicled how nascent bourgeois culture, with its hypocritical moral imperatives and capital accumulation exigencies, repressed the subversive spirit and basic livelihood of the peasantry. Rabelais was a utopian communist after Lefebvre's own heart; if party communism resembled Thomas More's *Utopia,* with its ordered, regimented island paradise, hermetically sealed off from anything that might contaminate it, Lefebvre's was a libertarian "Abbey of Thélème," with neither clocks nor walls. There, Rabelais urged "hypocrites and bigots, cynics and hungry lawyers" to "stay away"; there, laws and statutes weren't king but people's "own free will": "DO WHAT YOU WILL," proclaimed Rabelais, as he clinked glasses with a few old pals.[33] "Our Rabelais," writes Lefebvre, "had a utopia at once less immediately dangerous than More's, [yet] more beautiful and more seductive … a strange abbey, not a church but a fine library … an immense chateau."[34] Inside, everybody drank, sang and played harmonious music, spoke five or six languages, wrote "easy poetry" and "clear prose."[35] The Abbey of Thélème even seemed to anticipate the young Marx's radiant vision: a "communist utopia that opened itself resolutely towards the future. It proposed an image of man fully developed, in a free society. At the same time, Rabelais knows he's dreaming, because this society is headed towards terrible ordeals and chronic catastrophes."[36]

Lefebvre himself stepped into Rabelais's unfettered world of *excess,* affirming throughout his life his own free will in everyday life—for better or for worse. He willfully ignored abstinence and austerity, as well as a plebeian asceticism that informed a lot of his generation's visions of Marxism and communism. All the same, there was a flip side to Lefebvre's Rabelaisian nature, something not entirely positive; not least was his excessive publication (writing books at a rate that the most prolific wrote articles) and his excessive libido (like Rabelais's vagabond hero Panurge, extricated from all social and familial ties, Lefebvre seemed obsessed with a search for women and had a penchant for marriage). Indeed, Lefebvre's Rabelaisian excesses make his output effervescent and vital yet repetitive and overblown, like a drunk who repeats the same old joke to the same cronies every night at the bar. It was excess, too, that made much of Lefebvre's personal life chaotic, leaving ex-partners and ex-wives to pick up the pieces of a Lefebvrian personal liberty.

Still, excess became a redoubtable *political* force, and the most magical, supreme, and excessive event Rabelais documented—the peasant festival—enacted a joyous, primal kind of liberty that Lefebvre would never renounce, either personally or politically. He envisages the festival as a special, potentially modern form of Marxist praxis that could erupt on an urban street or in an alienated factory. The festival was a pure spontaneous moment, a popular "safety value", a catharsis for everyday passions and dreams, something both liberating and antithetical: to papal infallibility and Stalinist dogma, to Hitlerism and free-market earnestness, to bourgeois cant and born-again bullshit. Popular laughter existed outside the official sphere: it expressed idiom and a shadier, unofficial world, a reality more lawless and more free.

One of the most stirring instances of this was the *Fête des Fous* ("Feast of Fools"), celebrated across medieval France on New Year's Day. Festivities here were quasi-legal parodies of "official" ideology: masquerades and risqué dances, grotesque degradations of church rituals, unbridled gluttony and drunken orgies on the altar table, foolishness and folly run amok, laughter aimed at Christian dogma—at any dogma. These feasts were double-edged. On one hand, their roots were historical and steeped in past tradition, wore an ecclesiastical face, and got sanctioned by authorities. On the other hand, they looked toward the future, laughed and played, killed and gave birth at the same time, and recast the old into the new; they allowed nothing to perpetuate itself and reconnected people with both nature and human nature. As Lefebvre suggests (p. 57), "the celebration of order (terrestrial, thus social and cosmic) is equally the occasion of frenetic disorder." The fête situated itself at the decisive moment in the work cycle: planting, sowing, harvesting. Prudence and planning set the tone in the months preceding festival day, until all was unleashed: abundance and squandering underwrote several hours of total pleasure.

Laughter evoked—can still evoke—an interior kind of truth. It liberated not only from external censorship but also from all internal censorship. People became deeper, reclaimed their true selves, by lightening up. Laughter warded off fear: fear of the holy, fear of prohibitions, fear of the past and fear of the future, fear of power. It liberated—can still liberate—people from fear itself. Seriousness had an official tone, oppressed, frightened, bound, lied, and wore the mask of hypocrisy. It still does: we know this world all too well. (Or else the laughter of presidents exhibits *real* buffoonery, a little like the moronic *Ubu Roi* of Alfred Jarry, Rabelais's more modern successor. "Shittr," said Jarry's fictional cretin king, "by my green candle, let's go to war, since you're so keen on it!") But on festival day, all masks were dropped, all ideology exposed, all pretence pilloried. On festival day, another more immediate truth was heard, in frank and simple terms, amidst the laughter and the foolishness—because of the impropriety and parody. People literally drank and laughed away their fears. Laughter opened up people's eyes, posited the world anew in its most naive and soberest aspects.

In 1955, Lefebvre warned how we'd lost Rabelais's laughter. And in losing it, he said, we've lost a big part of our cultural heritage, even lost a weapon in our revolutionary arsenal. Lefebvre's study of Rabelais, by embracing festival, laughter, and the medieval sage as educator, evokes another instance of his "regressive–progressive" method: going backward, he suggests, helps us go forward and onward. For Lefebvre, the laugh of Rabelais bawled the song of innocence, not a song of deception, "a naïve life that sets its own laws upon solid principles, without struggling against itself nor without having to repress. … It's thus that the living humanism of Rabelais can serve the socialist humanist cause: by laughing."[37]

2

MOMENTS

A roll of the dice will never abolish chance.

—Stéphane Mallarmé,
Un coup de dés jamais n'abolira le hasard

Henri Lefebvre started the tumultuous decade of the 1960s laughing as a new man, if not quite a total man. He celebrated his sixtieth birthday with a new job (chair of sociology at the University of Strasbourg), a couple of new books (*Critique of Everyday Life—Volume 2* and *Introduction to Modernity*), and some new militant friends, younger friends—like Guy Debord and the Situationists—who'd ignite each other in the explosion of 1968. (By that point, Lefebvre was teaching at suburban Paris-Nanterre, where the initial spark of student discontent had been generated.) In 1961, he also inaugurated himself with a new status:

"turbo-Prof," a species of French academic who teaches in the provinces, who catches the *train à grande vitesse* (TGV) every Monday morning and Thursday afternoon, yet keeps a primary residence in the nation's capital. (Lefebvre lived at rue Rambuteau in the 3rd arrondissement, first in an apartment at number 24, and later at number 30. Both buildings were close to the old Les Halles market halls, architectural jewels destined to be demolished in 1969 to make way for the RER rapid computer train line—which would ironically speed to Nanterre. In 1977, the dreaded Pompidou Centre became Lefebvre's unwelcome, upscale neighbor. He could almost spit at it from his front window.)[1]

"Around 1960," Lefebvre reflects in *Everyday Life in the Modern World* (1968), "the situation became clearer." Everyday life was "no longer the no-man's land, the poor relation of specialized activities. In France and elsewhere, neo-capitalist leaders had become aware of the fact that colonies were more trouble than they were worth and there was a change of strategy; new vistas opened out such as investments in national territories and the organization of home trade."[2] The net result, Lefebvre thinks, was that "all areas outside the centers of political decision making and economic concentration of capital were considered as semi-colonies and exploited as such; these included the suburbs of cities, the countryside, zones of agricultural production and all outlying districts inhabited, needless to say, by employees, technicians and manual laborers; thus the state of the proletarian became generalized, leading to a blurring of class distinctions and ideological 'values.' "[3] Work life, private life, and leisure were "rationally" exploited, cut up, laid out, and put back together again, timetabled and monitored by the assorted bureaucracies, corporations, and technocracies.

Massive scientific and technological revolutions became a perverse inversion of—and substitute for—the social and political revolution that never materialized. That was like waiting for Godot. ("We'll hang ourselves tomorrow. [*Pause.*] Unless Godot comes. / And if he comes? / We'll be saved / … Well? Shall we go? / Yes, let's go. / *They do not move.*) And when Russian tanks rolled into Budapest in 1956, crushing Hungary's democracy movement, it publicly confirmed what Lefebvre privately already knew: the Soviet revolution had failed, had betrayed everyday life. And China's situation was uncertain and suspect. So "there was this gap," Lefebvre says, "and then the rise of a new social class, that of the technocrats. And then the advent of the world market—that is the world market after the period of industrial capitalism. This world market became an immense force with consequences even for the 'socialist' countries."[4] Moreover, the massive technological revolution was matched by equally massive processes of urbanization and modernization, which began transforming industries and environments everywhere, seemingly without limit, opening out new vistas while creating new voids, not Rabelaisian Abbeys of Thélème but new desert spaces, Alphavilles of the body and soul.

* * *

The *Critique of Everyday Life—Volume 2* (1961) not only revisited Lefebvre's old thesis a decade and a half on but also cut a swath—or "Cleared the Ground," as he put it—for a new platform of struggle, plotting a new revolutionary Northwest Passage within and beyond everyday life. In his one-hundred-page opener, Lefebvre seems to invent the 1960s in his own head, threading his way through "the labyrinthine complexities of the modern world."[5] Early on, he gives us a neat summation of his work to date. First, he'd reinstated a new Marxist agenda, a project both utopian and practical, based on the idea of a social praxis resolving contradictions and eliminating alienating divisions. Second, he'd grounded this agenda in everyday life, shedding light on what precisely revolution would change and could change, if anything. Third, he'd continue to monitor the "lags" between the real and the potential, the possible and the impossible, between "ethical patience" and "aesthetic irony." Volume 1, he claims, had hooked up everyday life with history and politics; now, "we must build a long-term policy on how to answer demands for a radical transformation of everyday life."[6]

Since 1947, the world had moved on; the economy was expanding, despite inherent crises, forever melting things into air, appropriating both external and internal nature, transforming social life into economic life, goods into needs, consumer whims into subliminal desires. Everyday life had been saturated with commodity logistics; corporate logos were set to become the semiotics of daily life they are today—a "semantic field" of ideological colonization. White-collar managers and industrial strategists, technocrats and bureaucrats began calling the shots, tallying work and family and social life with paradigms of order and efficiency. Low-grade alienation flourished through middlebrow affluence; in desolate suburbs and faraway New Towns, "lonely crowds" met "one-dimensional men." Everyday life, says Lefebvre, now reigned in its "chemically pure state"; social life more and more shrank into a decaffeinated and deerotized private life. Indeed, a "reprivatization of life" was in our midst, in tandem with a new round of capitalist modernity, which is intent on philosophizing life, converting it into speculative contemplation. "Predictable and expected," he writes, " 'globalization' is being achieved by a mode of withdrawal. In his armchair, the private man—who has even stopped seeking himself as a citizen—witnesses the universe without having a hold over it and without really wanting to. He looks at the world. He becomes globalized, but as an eye, purely and simply."[7]

Alienation of this sort likewise prompted scholarly reactions across the Atlantic. Sociologists David Riesman, Nathan Glazer, and Reuel Denney coined the name "lonely crowd," bemoaning a new kind of "other-directed" character, a uniformed mass-person who now slavishly follow the tyranny of majority, replacing centuries of American pioneering individualism and entrepreneurial "inner-directed" character traits.[8] Meanwhile, William H. Whyte highlighted how values of the corporate boardroom seeped into nonwork life. In exchange for security and high living standards, Americans voluntarily gave themselves over to "organization men," internalizing the latter's conformist principles, helping convert a business ethic into a general social ethic.[9] The product becomes self-fulfilling, Whyte suggested, a form of self-censoring and "togetherness" that's difficult to dislodge.

This line was also reiterated by Lefebvre's nearest radical peer over the ocean, the German émigré Herbert Marcuse, whose Hegelian–Freudian–Marxist *One Dimensional Man* saw a sinister high-tech "Total Administration" possessing the body and minds of everyday people, pacifying dissent, and instilling in them a delusional "happy consciousness."[10] For Marcuse, the Total Administration permeated all reality: it existed (exists?) in defense laboratories, in executive offices, in governments, in machines, in timekeepers and managers, in efficiency experts, in mass communications, in publicity agencies, in schools and universities. Through these consenting means, all opposition was thereby liquidated or else absorbed; all potential for sublimation, for converting sexual energies into political energies (and vice versa) was repressed and *desublimated.* The Reality Principle vanquished over the Pleasure Principle, convincing people that Reality was the *only* principle. Society had thus reclaimed even the space of imagination and dream.

"I met Marcuse several times," Lefebvre remarks in *Conversation avec Henri Lefebvre.* "We had some points of agreement on the critique of bourgeois society and one-dimensional man … but I didn't agree with him on the fact that one could change society by aesthetics. … According to Marcuse, industrial society, by its mode of social control, provokes a reductionism of possibilities for individuals and an integration (or disintegration) of the working class. The attack on the system can only come from an encounter between critical theory and a marginal substratum of outcasts and outsiders. But in May 1968 this attack took the form of a formidable working class general strike."[11] For Lefebvre, Marcuse's vision is tainted with closure and a pessimism that isn't so much reductive as restrictive, something wrenched out of everyday life. All revolt, in Marcuse's eyes, would come from those *outside* of the everyday: society's rejects and fugitives. As Lefebvre states in *The Explosion* (1968), "Marcuse's theory carries the thesis of 'reification' to its extreme conclusion and extends it from consciousness to the whole of reality. There is no question of refuting it. … Any movement within it is but illusion. The horizons are closed off. Only the desperate may attempt an assault. Herbert Marcuse *makes refutation impossible*. Irrefutable!"[12]

* * *

But no throw of the dice, for Lefebvre, can ever abolish chance, even if the game is rigged. No system of control can ever be total, Lefebvre maintains, can ever be without possibility, contingency, inconspicuous cracks, holes in the net, little shafts of light, and pockets of air. Lefebvre could never comprehend modern capitalism as seamless; his mind reveled in openness not closure; he was a butterfly not an inchworm.[13] Commodification and domination are real enough, he knew, yet they hadn't overwhelmed everything, not quite. There is always leakiness to culture and society, unforeseen circumstances buried within the everyday, immanent "moments" of prospective subversion. In this vein, the *moment* became his key revolutionary motif, signifying that all was not lost, that all could never be lost. Thus, in the final chapter of the *Critique of Everyday Life—Volume 2,* Lefebvre presents again his "Theory of Moments," first unveiled a few years prior in *La*

Somme et le Reste. "We will call 'Moment,' " he says, "*the attempt to achieve the total realization of a possibility*. Possibility offers itself; and it reveals itself. It is determined and consequently it is limited and partial. Therefore to wish to live it as a totality is to exhaust it as well as to fulfill it. The Moment wants to be freely total; it exhausts itself in the act of being lived."[14]

The "moment" assumed the same gravity for Lefebvre as white spaces between words did for the poet Stéphane Mallarmé. "The blank," the latter said, intervenes in the text to such a degree that it becomes part of the work itself. It becomes a secret door letting the reader enter. Once inside the reader can subvert each verse, rearrange its rhythm, reappropriate the poem as a covert author: "The text imposes itself," Mallarmé wrote, "in various places, near or far from the latent guiding thread, according to what seems to be the probable sense."[15] Mallarmé's poetry disrupted linear textual time much as Lefebvre's theory of moments sought to disrupt Henri Bergson's notion of linear real time—his *durée,* or duration. Creation, for Bergson, is like the flow of an arrow on a teleological trajectory. "The line [of the arrow] may be divided into as many parts as we wish," Bergson said, "of any length that we wish, and it will always be the same line."[16] Life itself, Bergson insisted, unfolds with similar temporality, and we comprehend ourselves in his unbroken, absolute time, not in space: "we perceive existence when we place ourselves in duration in order to go from that duration to moments, instead of starting from moments in order to bind them again and to construct duration."[17]

Lefebvre goes against the grain of time's arrow of progress, building a framework of historical duration from the standpoint of the moment—from, in other words, the exact opposite pole to Bergson's. Lefebvre hated Bergson's guts. In *La Somme et le Reste* (*Tome II,* p. 383), he writes, pulling no punches, "If, during this period [1924–26], there was a thinker for whom we (the young philosophers group) professed without hesitation the most utter contempt, it was Bergson. This feeble and formless thinker, his pseudo-concepts without definition, his theory of fluidity and continuity, his exaltation of pure internality, made us physically sick." Time, says Lefebvre, isn't just about evolution but *involution:* "The duration, far from defining itself solely as linear and punctuated by discontinuities, re-orientates itself like a curl of smoke or a spiral, a current in a whirlpool or a backwash."[18] The Lefebvrian moment, like Mallarmé's, was there between the lines, in a certain space, at a certain time. It disrupted linear duration, detonated it, dragged time off in a different, contingent direction, toward some unknown staging post. The moment is thus an opportunity to be seized and invented. It is both metaphorical and practical, palpable and impalpable, something intense and absolute, yet fleeting and relative, like sex, like the delirious climax of pure feeling, of pure immediacy, of being there and only there, like the moment of festival, or of revolution.

The moment was what Lefebvre on numerous occasions calls "the modality of presence." A moment, be it that of contemplation or struggle, love and play, rest and poetry, is never absolutely absolute or unique. "There are," he says, "a multiplicity of undefined instances," even though, in the plurality, a specific moment is "relatively privileged," relatively absolute, definable, and definitive, at least for a moment.[19] Each moment, accordingly, is a "partial totality" and "reflected and refracted a totality of global praxis," including the dialectical relations of society with itself and the relations of social man with nature. Moments become absolute—indeed, Lefebvre says, they have a *duty* to define themselves absolutely. They propose themselves as impossible. They wager for random winnings, "for the heady thrill of chance."[20] The entire life of a moment becomes a roll of the dice, a stack of chips at the casino of modern life.

"The *revolutionary* aspect of non-linear time," Lefebvre explains in *La Somme et le Reste* (*Tome I,* p. 236, original emphasis), "appeared to me a lot more essential than all others: radical discontinuities blurred into a theory that placed involution, or its dissolution, on the same plane as revolution. Ultimately, I conspired that the theory of moments, considered as a unique philosophy and ontology, might eliminate the idea of human historicity." The political moment, as Lefebvre wills it, is a pure and absolute act of contestation: a street demo or flying picket, a rent strike or a general strike. Streets would be the staging, and the drama might be epic or absurd or both, scripted by Brecht or Chaplin or Rabelais—who could tell? It's meant to be spontaneous, after all. Lefebvre points out how Hegel and Marx each emphasized the importance of the "moment." All dialectical movement progressed through different moments: moments of skeptical, negative consciousness defined history for Hegel; moments of contradictory unity defined and structured capitalism for Marx. All reality for both thinkers was momentary, transient, in motion, in fluid state, whether as an idea or as material reality.

Just as alienation reflected an *absence,* a dead moment empty of critical content, the Lefebvrian moment signified a *presence,* a fullness, alive and connected. Lefebvre's theory of moments implied a certain notion of liberty and passion. "For the old-fashioned romantic," he quips in *La Somme et le Reste,* "the fall of a leaf is a moment as significant as the fall of a state for a revolutionary."[21] Either way, whether for the romantic or for the revolutionary—or for the romantic revolutionary—a moment has a "certain specific duration." "Relatively durable," Lefebvre says, "it stands out from the continuum of transitories within the amorphous realm of the psyche." The moment "wants to endure. It cannot endure (at least, not for very long). Yet this inner contradiction gives it its intensity, which reaches crisis point when the inevitably of its own demise becomes apparent."[22] For a moment, "the instant of greatest importance is the instant of failure. The drama is situated within that instant of failure: it is the emergence from the everyday or collapse on failing to emerge, it is a caricature or a tragedy, a successful festival or a dubious ceremony."[23]

The spirit of past revolutions, replete with all their successes and failings, seems nearby: of 1789 and 1830; of 1848 and the 1871 Paris Commune; of 1917, 1949, and 1959; of the 1968 "Student Commune" (though Lefebvre wouldn't know it yet). Moments don't crop up anywhere, or at any time, at whim or by magic. The moment may be a marvel of the everyday, Lefebvre says, but it isn't a miracle. Indeed, the moment has its motives, and without those motives it wouldn't intervene in "the sad hinterland of everyday dullness" (p. 356). It is everyday life where possibility becomes apparent in "all its brute spontaneity and ambiguity. It is in the everyday that the inaugural decision is made by which the moment begins and opens out; this decision perceives a possibility, chooses it from among other possibilities, takes it in charge and becomes committed to it unreservedly" (p. 351). Everyday life, consequently, "is the native soil in which the moment germinates and takes root" (p. 357).

* * *

The Lefebvrian moment bore an uncanny resemblance to "the situation" of Guy Debord, the intense, bespectacled, freelance revolutionary whom Lefebvre befriended in 1957. Debord was thirty years Lefebvre's junior, a brilliant theorist and ruthless organizer, a poet and experimental filmmaker, the brainchild behind a militant crew of artists, poets, and students who hailed from France, Britain, Italy, Denmark, Belgium, and Holland. They'd banded together in a remote Italian village in July 1957, "in a state of semidrunkenness," to establish the so-called Situationist International (SI), an amalgam of hitherto disparate avant-garde organizations. The SI, which endured until 1972, was highly politicized in its intent to renew art—or, better, to "abolish" art, much as Marx sought to abolish philosophy—and to renew the action of art on life (and life on art). They were bored with art, bored with politicians, bored with the city, bored in the city. The city had become banal; art had become banal; politics had become banal—it still is. Everything needed changing: life needed changing, time and space needed changing, cities needed changing. Everybody was hypnotized by production and conveniences, by sewage systems, elevators, bathrooms, and washing machines. Presented with the choice between love and a garbage disposal unit, Debord once jeered, young people opted for a garbage disposal unit.

The Situationists, and Guy Debord notably, exerted a strange grip on Lefebvre. He began teaching fringe members at Strasbourg in the early 1960s, likely pages from his *Critique of Everyday Life,* and word spread fast; the SI, in turn, seemed to radicalize the aging professor, kept him on his toes, taught him a thing or two about praxis, forced him to up the ante in the classroom. Teachers and students both felt something brewing, gurgling within postwar culture and society, ready to erupt. Debord embodied the pure liberty Lefebvre admired, perhaps even envied.[24] The young Parisian who was neither student nor professor fascinated Lefebvre. Nobody knew how Debord got by; he had no job, didn't want a job. In fact, in 1953, he'd chalked on the wall of the rue de Seine a refrain that would become a sacred Situationist shibboleth: "*Ne Travaillez Jamais*"—"Never Work!" Later in life, Lefebvre recalled Debord (with his then-wife Michèle Bernstein) inhabiting "a kind of studio on rue Saint Martin, in a dark room, no lights at all." Not very far from his own rue Rambuteau apartment, it was "a miserable place, but at the same time a place where there was a great deal of strength and radiance in the thinking and the research."[25]

Lefebvre and Debord became acquainted through women. Bernstein's childhood friend, Évelyne Chastel, was Lefebvre's girlfriend—despite the big age gap. One day, both couples bumped into each other on a Parisian street not long after Lefebvre had quit the French Communist Party. Debord was very happy to finally meet the theorist whose books he'd read and much respected. "I remember marvelous moments with Guy," Lefebvre recalls in *Le Temps des Méprises,* "warm friendship, free of all mistrust and ambition."[26] He and Debord met and drank together, often talked all night, and engaged in "more than communication," Lefebvre admits, "a communion—which remains an extremely vivid memory."[27] Lefebvre was probably Debord's only *living* influence. Meanwhile, the young man steeped in Hegel and the early Marx, Cardinal de Retz and Lautréamont, charmed Lefebvre. "I remember very sharp, pointed discussions," Lefebvre says, "when Guy said that urbanism was becoming an ideology. He was absolutely right." Debord and the Situationists would stay a subject close to Lefebvre's heart. "One I care deeply about," he mused, years later. "It touches me in some ways very intimately because I knew them very well. I was close friends with them. The friendship lasted from 1957 to 1961 or '62, which is to say about five years. ... In the end, it was a love story that ended very, very badly."[28]

If Lefebvre's womanizing brought him and Debord together, it was the former's libido that eventually helped drive the two men apart. Debord held Lefebvre's Don Juan antics in low esteem, thought them comical and reprehensible, especially because the old prof–charmer dated women young enough to be his granddaughters, often getting them pregnant. One tale is recounted by Bernstein, whose friend Nicole Beaurain, a young student, was bearing Lefebvre's kid (after he'd split with Évelyne), a man "so old," Bernstein said, "and already several times a father." Bernstein, unsurprisingly, was dead against the pairing, and, according to Lefebvre, she and Debord sent an envoy, another of Bernstein's friends, Denise Cheyre, down to Navarrenx to persuade Nicole to abort. Lefebvre accused Debord of meddling; Debord apparently insulted Lefebvre on the phone. "I didn't see Guy mixing himself up in this affair," Bernstein remembered, years later. "It wasn't his style."[29]

During Lefebvre and Debord's early friendship, it was hard to know who influenced whom. Lefebvre may have lifted more from Debord and the Situationists than he cracked on. He and Debord were like a Faust–Mephistopheles pairing: Lefebvre, the old intellectual, fraternizing with the devilish powers of Debord, a darker figure, a man of the night uniting with Lefebvre's personality of the sun. In rescuing the sun from crucifixion, Lefebvre seemingly summoned up the spirit of darkness. Debord and the Situationists internalized the destructive–creative powers Lefebvre secretly harbored within himself; they were his catharsis incarnate, his kids of Dionysus. "We brought fuel to where the fire was," Debord explained in his film *In Girum Imus Nocte et Consumimur Igni*. "In this manner we enlisted definitively in the Devil's party—the 'historical evil' that leads existing conditions to their destruction, the 'bad side' that makes history by undermining all established satisfaction."[30]

In concert, Faust and Mephistopheles read Malcolm Lowry's *Under the Volcano,* about the doomed alcoholic antihero Geoffrey Fermin, supping not a few mescals themselves; Debord even helped organize Lefebvre's teaching schedule. Debord and Bernstein sometimes sojourned at Lefebvre's summerhouse in Navarrenx. And through Lefebvre, Debord met the young Belgian poet and free spirit Raoul Vaneigem, another avid Lefebvre reader who'd soon enter the Situationist fray. Around this time, too, Lefebvre discovered the Dutch utopian architect and planner Constant Nieuwenhuys and other anarchist "Provos" in Amsterdam, who later came to Paris and discovered Debord and his crew. Debord and Nieuwenhuys steadily nudged Lefebvre toward an interest in urbanism, which would soon hatch in *Introduction to Modernity.* "I went to Amsterdam to see what was going on," Lefebvre recalls. "There were Provos elected to the city council in Amsterdam. … Then, after that, it all fell apart. All this was part and parcel of the same thing. And after 1960 there was the great movement in urbanization … from that moment the historic city exploded peripherally, into suburbs, like what happened in Paris and all sorts of places."[31]

As with the Lefebvrian moment, "situations" were slippery, playful inventions and interventions, as much metaphorical as material. They were meant to be fleeting happenings, moving representations, the "sum of possibilities."[32] They'd be something lived but also "lived-beyond," full of immanent possibilities. Debord and the Situationists wanted to *construct* new situations, new life "concretely and deliberately constructed by the collective organization of a unitary ambience and a game of events." "New beauty," Debord proclaimed, "will be the beauty of situation."[33] Situations would be practical and active, designed to transform context by adding to the context, assaulting or parodying context, especially one where the status quo prevailed. What would emerge was a "unitary ensemble of behavior in time."

A vital trope here was *détournement,* or reversal and hijacking, which would scupper accepted bourgeois behavior and received ideas about places and people. Squatting, and building and street occupations are classic examples of *détournement,* as are graffiti and "free associative" expressionist art. All these actions would exaggerate, provoke, and contest. They'd turn things around, lampoon, plagiarize and parody, deconstruct and reconstruct ambience, unleash revolts inside one's head as well as out on the street with others. They'd force people to think and rethink what they once thought; often you'd not know whether to laugh or to cry. Either way, *détournement* couldn't be ignored: it was an instrument of propaganda, agitprop, an arousal of indignation, action that stimulated more action. They were a "negation and prelude," inspired by Lautréamont's *Poésis,* one of Debord's favorite works.

Debord meticulously studied Lefebvre's theory of moments. "At present," he told his friend André Frankin, in a letter dated February 14, 1960, "I am reading *La Somme et le Reste*. It is very interesting, and close to us—here I mean: the theory of moments."[34]

A week on (February 22, 1960), Debord wrote the same friend a long, detailed letter, analyzing Lefebvre's "theory of moments." (The correspondence would filter into Debord's critical article "Théorie des moments et construction des situations," which appeared in *Internationale Situationniste,* no. 4, 1960.) Debord's discussion is very technical and very serious: you sense the political stakes are high here. He thinks Lefebvre's moments are more durable, more precise, more pure than the Situationists's notion of situations, yet this might be a defect. Situations are less definitive, potentially richer, more open to mélange, which is good—except, says Debord, how can "one characterize a situation": Where does it begin, and where does it end? At what point, and where, does it become a different situation?[35] Could the lack of specificity hamper effective praxis? Could too much specificity turn a situation into a moment? What, he asks, is a unique moment (or situation), and what is an ephemeral one?

The chief fault of Lefebvre, according to Debord, a fault that perhaps anticipates—or provokes—Lefebvre's "spatial turn" to come, is that his moment is "first of all temporal, a zone of temporalization. The situation (closely articulated to place) … is completely spatiotemporal." Situations are much more spatial, Debord thinks, and much more urban in orientation than the Lefebvrian moment. "In the end," Debord told Frankin, "for resuming the problem of an encounter between the theory of moments and an operational theory of the construction of situations, we would need to pose these questions: what mix? What interaction? Lefebvre is right in at least this: the moment tends toward the absolute, and devours itself in that absolute. It is, *at the same time,* a proclamation of the absolute and a consciousness of its passage."[36]

* * *

Lefebvre saw in the Situationists the germ of a "new romanticism," a brazenly utopian response to the problems of technological culture and industrial civilization. In them, he spotted a renewal of both classical and modern romanticism fighting back against the boredom and rationality of a bourgeois modernity run amok, updating the project the French novelist Stendhal announced in the 1820s: "At last," Stendhal wrote in *Racine and Shakespeare* (1823), "the great day will come when the youth of France will awake; this noble youth will be amazed to realize how long and how seriously it has been applauding such colossal inanities."[37] "It requires courage to be a romantic," Stendhal claimed, "because one must take a chance." Stendhal's critical text, insists Lefebvre 140 years on, is of "vital importance," going far beyond the limits of literary criticism as such, because it attacked the values of the Restoration (the moral order, imitation of the rich and powerful, pedantry) and spearheaded an alternative direction to French social and political life from 1825 onward. "In 1961," Lefebvre asks, "are we heading towards an analogous renewal in literature, and beyond literature? That is the question we intend to examine—inconclusively, no doubt—using Stendhal's book as our starting point, and attempting to reason, via similarities and differences between his times and ours."[38]

In the "Twelfth Prelude" of *Introduction to Modernity,* Lefebvre revivifies the spirit of Stendhal as the antecedent of Guy Debord, mocking and jousting with his Situationist friends as they sometimes mocked and jousted with him. The dense, 150-page *dénouement,* which expressed guarded, critical admiration for the Situationists as "a youth movement," marked the centerpiece of Lefebvre's 1962 inquiry into the "more and more brutal, more rapid, more noisy" march of the modern world. Written as a series of "preludes," evoking some mix of playful musical motif—the book's wish, he says, is to be understood in "the mind's ear"—unfinished, fragmentary explorations, and, perhaps, a testament of what's to come, Lefebvre's thesis follows a deliberately winding path, full of bewildering twists and turns and shifts of melody and tonality.

In the Situationists he recognizes a new avant-garde generation, different from the "Lost" or "Beat" generations, angrier and more realistic than the Surrealists, and less angry and more humorous than Lefebvre's generation of communists. "The most brilliant Situationists," Lefebvre suggests, "are exploring and testing out a kind of *lived utopianism,* by seeking a consciousness and a constructive activity which will be *disalienating,* in contradistinction to the alienated structures and alienating situations which are rife within 'modernity.' "[39] And yet, he cautions, we mustn't accredit too much to the Situationists and their ilk. After all, youth is an age, not a social class, and thus they cannot fulfill any "historical mission." "Yes, because it is an avant-garde, it scours the future. It marches in the vanguard, scanning and prefiguring the horizon;" but, no, it cannot change life alone, not without soliciting the help of an organized working class. Transforming hypothetical exploration into a political program, into an applicable plan, plainly requires real participation: real unified practice.[40]

Lefebvre's conclusion to *Introduction to Modernity*—which seems to be introducing *his* notion of a modernity to come—is simply that there are indications of a "new attitude" drifting in the breeze: revolts, acts of insubordination, protests, abstentions, and rebellions are, he says, there to be seen and felt; Stendhal is a man of the late twentieth century. Stendhal took the *pleasure principle* as his opening gambit, and "in 1961," Lefebvre goads, "can we bring the pleasure principle back as a foundation, a starting point, and believe in the creative virtues of pleasure?"[41]—a question we still need to confront today. What the romantics saw around them then, and what the "new romantics" see around them now, is a world no longer governed by constraints: "in the name of lived experience," Lefebvre notes (p. 291), they reject scientism and positivism and find their place in a chaos of contradictory feelings in a society riven by upheavals, convulsions, and irresolvable conflicts. "This offered their extravagant subjectivities a total—or apparently total—adventure."

Stendhal's romanticism affirmed disparate elements of society: "women, young people, political rebels, exiles, intellectuals, who dabbled in deviant experiments (eroticism, alcohol, hashish), half-crazed debauchees, drunks, misfits, successive and abortive geniuses, *arrivistes,* Parisian dandies and provincial snobs."[42] This ragged, motley array of people attempted to live out, within everyday bourgeois society, their ideal solutions to bourgeois society, challenging its moral order, surviving in its core, "like a maggot in a fruit," trying to eat their way out from the inside. They sought to *reinvent* the world. And using all their powers of symbolism, imagination, and fiction, a new subjectivity was born, a new lived experience conceived; outrageous fantasy succeeded in shaping grubby reality. Could, wonders Lefebvre, a "new romanticism" do the same in the 1960s? Could a "new" new romanticism do it at the beginning of the twenty-first century? And who are the "maggots" eating their way out from the inside of our rotten society?

3

Spontaneity

Bestir yourself!—Ah, for us science doesn't go fast enough!

—**Arthur Rimbaud,** *Une saison en enfer*

Not too long after the dramatic irruptive moments on the streets of Seattle, protesting the World Trade Organization's summit, I taught a class on Marxist urbanism at a Massachusetts liberal arts college. One of the key texts I'd chosen was Lefebvre's *The Explosion,* written only months after the even more dramatic student uprisings of May 1968. The images of street fighting and police heavy-handedness, circa *fin de millénaire,* surprised many pundits—radicals and conservatives alike—and I remember having little inkling of what Lefebvre's text, dictated almost as cars blazed in central Paris, could tell us as smoke still smouldered in downtown Seattle.

Lefebvre's window on the events of 1968 was particularly fascinating, given he'd had a foot in each camp: the ex-communist, expelled from the party for "ideological deviations," nonetheless remained a socialist true believer and a maverick fellow traveler; meanwhile, a lot of active participants in the demos and occupations, like Nanterre sociology major and *Rouge et Noir* militant Daniel Cohn-Bendit, had read and listened to Lefebvre and were somehow putting his lectures into practice. "Oh, he was a wonderful lecturer," Cohn-Bendit told me, recently. "He would seduce everybody, just talk, telling anecdotes; he loved to talk and everybody loved his classes." A twenty-one-year-old Cohn-Bendit, a prominent student agitator and spokesperson, was among the two-thousand-odd students who followed Lefebvre's class on modernity and everyday life in Amphithéâtre B at Nanterre, 1966–67. "I didn't really know him personally," admitted Cohn-Bendit. "I was only one of many students in the audience. But his ideas on cultural-revolution in everyday life, and on offering a different version of Marxism, influenced the 'Movement of March 22nd.' "[1]

On that notorious March day, assorted Situationists, young communists, Trotskyists, anarchists, and Maoists invaded Nanterre's administration building and began occupying it. Posters went up and slogans were scribbled on the walls of Nanterre in peripheral west Paris and soon at the Sorbonne in the Latin Quarter: "TAKE YOUR DESIRES FOR REALITY," "NEVER WORK," "BOREDOM IS COUNTERREVOLUTIONARY," "TRADE UNIONS ARE BROTHELS," "PROFESSORS, YOU MAKE US GROW OLD," "IF YOU RUN INTO A COP, SMASH HIS FACE IN." In early May, "the March 22 Movement" met with UNEF (the French National Student Union) at the Sorbonne. The authorities tried to break up the meeting but instead only unleashed its latent power. On May 6 and 7, a huge student demonstration took over the Boulevard Saint Michel and thoroughfares near rue Gay-Lussac; protesters overturned cars and set them alight, dispatched

Molotov cocktails, and manned the barricades. On May 13, there was a one-day general strike; "student–worker" solidarity suddenly looked possible, against the French Communist Party's and general worker's union's odds. By May 20, strikes and occupations became contagious. Nationwide, around ten million workers downed tools and froze assembly lines. France seemed on the precipice of revolution; a festival of people was glimpsed, briefly.

Lefebvre's double allegiance with the students and the working class meant his Marxist take on May 1968 was at once orthodox and heterodox, rooted in the "objective conditions" of French postwar society, conditions expressive of economic contradictions and crisis tendencies in long waves of growth; on the other hand, he was equally sympathetic to the "specific" and "subjective" grievances of the youth: their alienation, their hatred of institutions, their loathing of the admen and technocrats plotting to commodify the world. (They were also voicing discontent over an all-too-persistent theme: an illegal war in a far-off place perpetrated by American military might.) In essence, the wily Lefebvre wanted to highlight what was simultaneously general and specific about this latest "French Revolution," what was objective and subjective, structural and superstructural, old and new in the situation. He wanted to grasp everything dialectically and explain things in their totality. He yearned, above all, for young and old progressives to dialogue around theory and action.

When I first read *The Explosion* in the late 1980s, it hadn't turned me on much. Doubtless the prevailing political climate hardly helped. After all, my friends and I, like much of the British left, were then afflicted with New Right blues, or were languid with the melancholy of postmodernism. A decade or more on, a few things had changed, some for the better! To begin with, the din around postmodernism had subsided: nowadays, the intellectual left isn't so much bothered about deconstructing Los Angeles's Bonaventure Hotel as a postmodern hyperspace as it is supporting the Hotel and Restaurant Employees' union reconstruct its rank and file there. Furthermore, a "post-Seattle" era was suddenly in our midst, with a many-striped multitude of foes now confronting corporate globalization and neoliberalism. Young people were out on the street again, and direct action was alive and apparently well, growing in strength. So Lefebvre's *The Explosion* sounded fresh again, and his insights sharp. Rereading chapters titled "Contestation, Spontaneity, Violence" struck as amazingly salient and suggestive for figuring out our own current situation. Here its lessons are two pronged and double edged, just as they were in 1968: *The Explosion* issues words of wisdom about critical analysis and radical tactics and duly throws down the gauntlet to both the New Left and what we might now call the "*new* New Left."

These days the New Left consists of those who came of age during the 1960s civil rights and antiwar movements, the youth of Lefebvre's time. They were once yippies and hippies and SDSers (Students for a Democratic Society) but are now the gray-haired and gray-bearded "used" Left, an assorted coterie of still-radical tenured professors, public school teachers, writers and intellectuals, and dedicated subscribers of *Monthly Review, Dissent,* and *The Nation.* The new New Left, on the other hand, coheres around members of the United Students against Sweatshops, young college kids launching consumer boycotts of campus garb made by toiling third world below-minimum-wage employees; others are straight out of college, ripe for high-paying jobs in the business world yet have rejected the whole corporate bit. Instead, they're unofficial lieutenants in autonomous organizations like Global Exchange and the Ruckus Society, footloose campaigners against the World Bank, the International Monetary Fund, and global trade inequities. Others are environmentalists with Friends of the Earth, Greenpeace, the Sierra Club, and the Rainforest Action Network. Still more are graduates of the Anti-Apartheid and Latin American democracy movements, or black-masked anarchists and various free spirits. All, however, are more likely to root for the Zapatistas than for Karl Marx. The generational rift between these two factions is apparent, as are their organization platforms and ideological bases.

In such a context, Lefebvre shines as somebody who brought—can still bring—together older socialists and younger protesters to analyze the same problematic and to act on the street. The issues he devoted himself toward haven't, alas, been resolved: changing life, changing society, the links between theory and praxis, between spontaneity and planning, between attack and defense. Lefebvre addressed these questions fifty years ago, and he can continue to help ferment the kind of oppositional lingua franca needed today, especially to move along resistance against neoliberalism and neoconservatism. Lefebvre thrived from creating new ideas and fresh ways of seeing and reinventing himself. Each reinvention built on an already accomplished body of work, yet took it further, propelled it onward; sometimes it tore it down, set it ablaze; frequently his notions combusted spontaneously. He was animated by the thought of "explosion," by something abrupt and sudden, by an event or practice unforeseen and unplanned. Indeed, explosive metaphors are writ large in Lefebvre's œuvre: he reveled in "detonation," in blowing things up, in stirring up magic potions that fizzle and create bubbles. The metaphor equally says a lot about his own explosive and impulsive character, about why he was and remains a dangerous thinker.

* * *

In the thirty years prior to the 1960s, Lefebvre believed radicalism all but extinct. Economic growth, material affluence, a world war and a cold war had destroyed, absorbed, bought off, and won over many intellectuals of his generation. Ghettoized or brainwashed, they either died off or killed themselves off, lost themselves or found themselves fraternizing with the mainstream, becoming the self-same bureaucrats and technocrats Lefebvre detested, propping up the institutions of modern power he critiqued—anonymous and depersonalized, clinical and Kafkaesque sorts of power. (Lefebvre knew he needed to develop younger friendships, if only to ensure he wasn't another sad victim.) And yet, "worldwide," he acknowledged on the cusp of revolt, with his finger typically on the pulse, "avant-gardes are forming again, and making their voices heard. It is an observable fact. … They are perfectly convinced that we are all caught up in a gigantic stupidity, a colossal, dreary, pedantic ugliness, which stands victorious over the corpses of spontaneity, taste and lucidity."[2]

In May 1968, students and workers at last began to realize, as they did in 1999, the gigantic stupidity they were caught up in. And in its taste for spontaneity and lucidity, as well as a desire to advance action and explain its intent, *The Explosion* sought to steer a dialectical path between the rationality of theory and the irrationality of action. Lefebvre tries to deal with the slippage between the two, between lucidity and spontaneity, recoupling thinking and acting within an explicitly political analysis, an analysis that opens up the horizon of possible alternatives. "Events," he insists at the start of the text, "belie forecasts."[3] Who, for instance, could have predicted with any certainty the turbulent Maydays in Paris or those of Seattle in November and December 1999? "To the extent that events are historic," he says, "they upset calculations. They may even overturn strategies that provided for their possible occurrence. Because of their conjunctural nature, events upset the structures which made them possible" (p. 7). As such, events are always original.

Nevertheless, original events always get reabsorbed into a "general situation," and their "particularities in no way exclude analyses, references, repetitions, and fresh starts" (p. 7). Nothing "is absolutely virginal, not even the violence which considers itself 'pure.' " So eruptive events are both unique and general, and are rude awakenings for those who show contempt for history or are preoccupied with stability. Eruptive events, Lefebvre says, in words that could easily have been directed at our current neoliberal bigwigs, "pull thinkers out of their comfortable seats and plunge them headlong into a wave of contradictions. Those who are obsessed with stability lose their smiling confidence and good humor" (p. 8). The big question that follows is, "What is new and what is certain in the midst of uncertainty?"

Some things change, others don't. In 1968, like today, Lefebvre recognizes the boredom (*ennui*) associated with Marxism, at least on the left. (Revulsion still prevails on the right.) Then, like now, history was apparently propelled by technology not by class struggle; the main dilemma was (is) no longer control of development but the "technical programming of the fruits of technology" (p. 10)—as Alan Greenspan reiterated throughout the 1990s. Then, as now, alienation was said to have disappeared in a society of abundance, leisure, and consumption. In 1968, French life was ruled by a technocracy and "monopoly capital" who tried to "deideologize" society, yet whose grip on that society was loosening. The older generation had previously wanted in, had demanded consumer goods, increased wages, refrigerators, and automobiles; the younger '68 generation wanted out, demanded something more—a bit like today's Seattle generation—asking what price the growth, what cost the material wealth?

Then, as now, a complex intermingling of cultural, political, and economic forces prevailed. Then, as now, there was a mixture of old and new contradictions. Of course, the basic class contradiction between private ownership of the means of production and the social character of productive labor, considered primary by Marx, remained unresolved in 1968—and still does. But ownership of these productive forces was in 1968, like now, no longer the same as in Marx's day. What had happened instead, Lefebvre thought in 1968, presciently, is a newer contradiction: the growth of "the entire complex of organizations and institutions engaged in management and decision-making. They are superimposed on the economic organizations proper, and constitute the foundation and instrument of what is called Power. They appear to constitute a system. The term 'capitalist system' has not lost its meaning in the century that has elapsed since the appearance in 1867 of Volume 1 of *Capital*. Far from it. Its meaning has become more precise. It has become clearly and distinctly political" (pp. 14–15).

Thus, on one hand, nations and nationality have been engulfed by economic factors and commodity dictates, pretty much as the *Manifesto* prophesied; on the other hand, Marx clearly overstated bourgeois commitment to "free trade," to its tearing down of every barrier to production and exchange. Indeed, the most powerful members of this class have collectively devised all sorts of regulatory (and deregulatory) devices to politically finagle and actively restrict, manipulate, and control certain markets (as well as the world market), establishing new superstate and suprastate authorities, new gigantic executive committees for managing the common affairs of the whole bourgeoisie[4]—hence an ever-growing list of organizations, trade agreements, and acronyms, bizarrely touting the virtues of free trade, of a neoliberalism without tears. Consequently, Lefebvre is right to suggest Marx "could not anticipate the flexibility and adaptive powers of these relations, and this in spite of his stipulation that capitalism had inherent limits" (p. 19). Nor, moreover, could Marx ever have foreseen "that critical and revolutionary Marxism would be transformed into the ideological superstructure of socialist countries" (p. 19). Neither could he foresee exactly where, and when, any radical contestation of this capitalist executive committee would flare up.

* * *

Marx never really elaborated a theory of "contestation," so in *The Explosion* Lefebvre lends a hand. What is crucial about contestation, Lefebvre believes, is "its aim to link economic factors (including economic demands) with politics" (p. 65). Contestation names names, points fingers, has institutions and men merge, makes abstractions real, and is one way "subjects" express themselves, ceasing to be "objects" of institutional will and economic capital. Contestation, Lefebvre says, "replaces the social and political mediations by which the demands were raised to an all-inclusive political level" (p. 65). In other words, contestation blooms because activists and contesters know, for certain, that capitalist representative "democracy" is a crock of shit. Contestation smacks as a *refusal* to be co-opted, a "refusal to be integrated." Integration symbolizes cowardice, and its rejection shows "an awareness of what integration entails with respect to humiliation and dissociation" (p. 67). Contestation is "born from negation and has a negative character; it is essentially radical" (p. 67). It "brings to light its hidden origins; and it surges from the depths to the political summits, which it also illuminates in rejecting them" (p. 67). It rejects passivity, fosters participation, arises out of a latent institutional crisis, transforming it into "an open crisis which challenges hierarchies, centers of power" (p. 68). Contestation "obstructs and undermines a rationality prematurely identified with the real and the possible" (p. 68) and pillories the complacency of institutional wishful thinking, especially ideologies of TINA—There Is No Alternative.

At the same time, contestation—the AFL–CIO (American Federation of Labor) might want to take note—"surges beyond the gap that lies between the realm of limited economic trade-union demands and the realm of politics, by rejecting the specialized political activity of political machines" (p. 68). In rejecting narrow economic demands, "contestation reaches the level of politics by a dialectical process that reflects its own style: critical and theoretical contestation, contesting praxis, and the theoretical examination of this process" (p. 69). Contestation "contemptuously and unequivocally rejects the ideology which views the passive act of consumption as conducive to happiness, and the purely visual preoccupation with pure spectacle as conducive to pleasure." "What does contestation seek to substitute for this ideology?" asks Lefebvre. "Activity," he answers, a "participation that is effective, continuous, permanent—participation which is both institutive and constitutive" (p. 68).

Contestation exposes "lags": lags between the people and the political process, lags between reality and possibility (the former always lags behind the latter), lags between consciousness and consciousness of consciousness itself. Contestation can help reality no longer lag behind dream. Frequently, Lefebvre maintains, contestation flares up spontaneously, and this can be a prodigiously creative force. In fact, contestation thrives off spontaneity, "has the outlook and limits of spontaneity" (p. 69). But Lefebvre recognizes its ambivalence and knows there's no such thing as "absolute" spontaneity anyway, as it erupts out of prior conditions and is never purely "savage." (Even the Direct Action Network [DAN], a conglomerate of grassroots groups who were most active in Seattle's downtown battles, had painstakingly planned through the Internet its street maneuvering months prior. A lot of their spontaneity actually arose in response to police heavy-handedness.)

The debate about spontaneity has a long and checkered history within socialism, having brought Rosa Luxemburg to blows with Lenin in 1904—to say nothing about dividing Marx and anarchist Mikhail Bakunin within the First International Working Men's Association (1864–76). Lenin belittled spontaneity, insisted it was a "subjective element" that couldn't congeal into a fully blown "objective factor." In *One Step Forward, Two Steps Back,* he said the "spontaneous development of the workers' movement leads precisely to its subordination to bourgeois ideology."[5] He reckoned a "socialist consciousness" could be brought to the people only from the outside. By itself, the working class is capable only of a restrictive, "pure-and-simple trade union consciousness." As a result, the working class needed a party, led by an elite vanguard, by dedicated intellectuals who would make revolution their calling, who would purge the movement of its spontaneity, dictate a tight, tactical program of action, especially "to rebellious students ... to discontented religious sectaries, to indignant school teachers, etc."

The Marxist–Leninist campaign against spontaneity, Lefebvre laments, has "been waged in the name of science, in the name of insurrection viewed as a technique, and in the name of organization" (p. 69). This had a catastrophic effect on looser, populist protesting, throwing the baby out with the bathwater. Indeed, certain strains of Marxism followed Lenin's edict that spontaneity was devoid of value, that it was essentially irrational. Spontaneity lacked the military discipline Lenin wanted, lacked his centralist take on organization, regressed into "tailism," with the tail wagging the dog, the masses steering the party, and a "slavish kowtowing before spontaneity."

Lefebvre's humanist Marxism bonds with Luxemburg's, mirroring Louis Althusser's antihumanist bonding with Lenin. (Althusser's Leninist-inspired *Reading Capital* appeared one year after *The Explosion*.) While Lefebvre's loose, energetic, rapid-fire formulations and spontaneous outpourings attracted student–militant readers, the clinical rigor and paired-down style of arch-Leninist Althusser likewise had appeal (especially in the post-'68 period when street spontaneity quieted). What Lefebvre articulated in weighty tomes, stretching for hundreds of playful pages, Althusser laid down solid in a chapter. The tight, disciplined, tactical theoretical and practical program that Lenin preached underwrote Althusser's best texts like *For Marx, Reading Capital,* and *Lenin and Philosophy,* where he constructed a "scientific" Marxist theory, grounded in concrete concepts, a veritable analytical tool with coherence and form. When the fighting stopped, and when people came up for air during the 1970s, Althusser's ideas thus cornered an ever-growing radical niche.

Luxemburg, however, like Lefebvre, has no truck with Lenin's "ultra-centralist tendency," rejecting his contempt for nonaligned working-class activism, for the "objectivity" of the party that Althusser equally underscored. Different progressive and working-class federations, Luxemburg wrote in *The Russian Revolution, and Leninism or Marxism?* needed a "liberty of action."[6] That way they could better "develop their revolutionary initiative and ... utilize all the resources of a situation." Lenin's line was "full of the sterile spirit of overseer. It is not a positive and creative spirit." Luxemburg is more generous, more sensitive to the ups and downs of struggle, in the course of which an organization emanates and grows, unpredictably pell-mell. Social democracy, she said, isn't just "invented"; it is "the product of a series of great creative acts of the often spontaneous class struggle seeking its way forward."

Of course, a movement might not immediately recognize itself within this class struggle, given people become *aware* of themselves objectively, as members of the working class, during the course of struggle. They define themselves through their opposite, through encountering a "ruling class," their other, people who are different from them, who have power and wealth and authority and whose interests are different from theirs, against theirs somehow. Class becomes *acknowledged* en route—not *a priori*—through a struggle for recognition, as Hegel would have said. Sometimes this could be misrecognition, too. There aren't any precisely prescribed sets of revolutionary tactics, no tactical recipe books. In fact, "the erection of an air-tight partition between the class-conscious nucleus of the proletariat already in the party and its immediate popular environment" is, for Luxemburg, mindlessly sectarian. The unconscious comes forth before the conscious; the movement, she said, advances "spontaneously by leaps and bounds. To attempt to bind the initiative … to surround it with barbed wire, is to render it incapable of accomplishing the tremendous tasks of the hour."

Lefebvre and Luxemburg should be on the reading lists of antiglobalizers everywhere; ditto for every Marxist. "Killing a spontaneous ideology, instead of trying to understand it and guide it toward a practice which may overcome it at the right moment—neither too early nor too late—that," Lefebvre maintains, "is a mark of dogmatism" (p. 70). Without spontaneity nothing happens, nothing progresses. "Power therefore regards spontaneity as the enemy" (p. 70). Always, spontaneity expresses itself in the street, the authentic arena of Lefebvre's Marxist politics, where it can spawn within everyday life, even transform everyday life, be festive, an intense Rabelaisian moment of everyday life, an emotional release.[7]

The street is an arena of society not completely occupied by institutions. Institutions fear the street: they try to cordon it off, try to repress street spontaneity, try to separate different factions of protesters in the street, quelling the apparent disorder, seeking to reaffirm order, in the name of the law. From street level, from below, contestation can spread to institutional areas, above. Spontaneous contestation can unveil power, bring it out in the open, out of its mirrored-glass offices, its black-car motorcades, its private country clubs, its conference rooms—sometimes it doesn't even let power into its conference rooms! Since Seattle, streets have become explicitly politicized, filling in the void left by institutional politics. In the streets, globalization is brought home to roost, somewhere. Therein lies the strength of spontaneous street contestation; therein lies its weakness: the weakness of localism, of symbolism, of "partial practice," of nihilism.

And yet, the explosion of street politics and spontaneity in Seattle, in Washington, in New York, in Davos, Switzerland (where the World Economic Forum meets annually), as well as in Quebec City (April 2001), where tear gas and water cannons met those protesting the Free Trade Area of the Americas talks, has led to the rebirth within radicalism of the phenomenon of violence. Violence is connected with spontaneity and with contestation—"with forces that are in search of orientation and can exist only by expressing themselves" (p. 72). Thirty thousand protesters expressed contestation by piling into Quebec where thirty-four heads of state gathered to talk about a "free trade" bloc for the Americas, a sort of NAFTA on steroids. Eight hundred million people would be drawn into its neoliberal remit, spanning Alaska to Argentina. "SMASH CAPITALISM!" one oppositional graffito read; "FREEDOM CAN'T BE BOUGHT!" said another. Cheerleaders, using bullhorns, sang "WELCOME TO THE CARNIVAL AGAINST CAPITALISM!" Gray-haired activists linked arms with their green-haired counterparts, and as well as marching in the street, chanting, and singing, they organized their very own "Peoples' Summit," with its counterglobalization manifesto, a grassroots version. Surrounding the venue was a giant chain-link fence, a security zone, keeping demonstrators strictly off-limits. "Wall of shame" became its nickname, before rabble-rousers tore it down.

Violence, for Lefebvre, is unavoidable in radical struggle. Breaking things up, making nonsense out of meaning (and meaning out of nonsense), throwing bricks through Starbuck's windows, driving tractors into McDonald's, burning cars, daubing graffiti on walls—all are justifiable responses to state repression and corporate injustice, to the "latent violence" of power. Hence they are legitimate forms of "counterviolence." In this sense, violence expresses what Lefebvre calls a "lag" between "peaceful coexistence" and "stagnating social relations," symptomatic of "new contradictions super-imposed on older contradictions that were veiled, blurred, reduced, but never resolved" (pp. 72–73). Lefebvre sees a certain political purchase in slightly mad destructive behavior, in senseless acts of beauty—so long as they don't degenerate into "the ontology of unconditional spontaneity," into "the metaphysics of violence" (pp. 73–74). Reliance only on violence, he concludes, leads to a "rebirth of a tragic consciousness" (p. 74), antithetical to the dialectic of becoming. Consequently, serious concern with contestation, spontaneity, and violence requires at the same time a serious delineation of spontaneity and violence. Yet this needs to be done in the name of theory, "which pure spontaneity tends to ignore" (p. 74).

* * *

Civic commotion to corporate promotion faces a predictable ideological barrage from mainstream media, from free-trade pundits, experts, consultants, business school professors, and "objective" economists—from those technocrats Lefebvre would christen "*cybernanthropes*." As ever, protesters are denounced as idiotic, juvenile, naive: listen up, wise up, and grow up. There is no alternative. Notwithstanding, "childish" pranks refuse to let up. "Immature" young people can still teach grown-ups a thing or two about mature life and politics. Even the sixty-something Lefebvre knew as much. He knew that maturity often spelled certitude, and certitude frequently translated into dogmatism; it tended to move from the relative to the absolute. On the other hand, incertitude spelled nihilism, lurched toward absolute violence, to a lot of people getting hurt, especially young people. Lefebvre frames the paradox thus: "Spontaneity acts like the elements: it occupies whatever empty space it can find, and sometimes it devastates this space. Thought offers another space, sometimes in vain; and other forms, sometimes to no avail" (p. 52).

Lefebvre wanted to stake out a position somewhere in between, somewhere that had a "unity of knowledge," retained "political awareness" and "theoretical understanding," and expressed "the scope and orientation of revolutionary truth" (p. 154). (He'd label it "cultivated spontaneity" in *The Survival of Capitalism*.)[8] It would center on concrete problems that are both practical and theoretical and would require at once sobriety and exuberance, diligent theory and mad raving ideals. It meant, too, an "unceasing critical analysis of absolute politics and the ideologies elaborated by specialized political machines" (p. 154). It was neither dogmatism nor nihilism but something else entirely, something Lefebvre ironically labels a "Third Way" (pp. 156–57). In no way should we confuse this with the closet neoliberalism of Giddensian "Third Wayers."[9] Instead, Lefebvre's Marxist Third Way keeps intact the notion that politics can be romantic, that the future can be different, that *we can still believe in the future*. As such, he warned long ago that the "centralized state is going to take charge of the forces that reject and, in essence, contest it. It will attempt this while at the same time forbidding contestation" (p. 52).

Contestation and struggle, transgression and creation are thus nonnegotiable Lefebvrian pairings. They go together like chalk and cheese. "Transgression," he says, "without prior project, pursues its work. It leaps over boundaries, liberates, wipes out limits" (p. 118). Perhaps most precious of all, as the state and ruling classes forbid protest, is that transgression marks "the explosion of unfettered speech" (p. 119). The transgressions of May 1968, as well as their new millennium counterparts, took and take "a devastating revenge on the constraints of written language. Speech manifests itself as a primary freedom"—we might say almost *primal freedom*. "In this verbal delirium, there unfolded a vast psychodrama, or rather a vast social therapy, an ideological cure for intellectuals and non-intellectuals, who finally met. All this speech had to be expressed for the event to exist and leave traces" (p. 119).

When protest is banned, outlawed, silenced, or pilloried in the press, contestation "will change into agitation and spectacle, and this spectacle will change into spectacular agitation" (p. 52).

"Spectacular agitation" has already been glimpsed, has already erupted on our streets, coalescing around many different agendas, voiced by many different groups, pitched at many different scales: canceling third world debt, banning child and sweatshop labor, ridding cars from our cities, keeping city life vital, saving turtles, shutting down the World Trade Organization and International Monetary Fund, taming unfettered globalization, changing the world, and changing life. Participation has shown its muscle: people have joined hands, especially as the batons flail and the tear gas flows. A reenergized militancy and spontaneity has reared its head. Its contestation has posed unflinching questions while it's grappled for answers. It has shown an amazing capacity to politicize people, especially young people, those disgruntled with ballet-box posturing and Bush banalities, people who care about our fragile democracy and our sacked society.

Some protagonists, like Global Exchange, a San Francisco–based human rights organization, comprise nomadic gadflies, young activists who travel up and down America, living in trailers and pickup trucks. They spread the anticorporate word at hitherto unprecedented decibels, mixing painstaking planning with spontaneous militancy, clearheaded analysis with touchy-feely utopianism. Indeed, their whole ontological raison d'être is *organizing:* politicking and proselytizing, conducting teach-ins and speak-outs, staging demos and boycotts, and masterminding blitzes, everywhere. Their ideas and ideals fill the gaping void that capitalist consumerism bequeaths young, intelligent people today. Global Exchange is also a prime mover in the umbrella group, DAN, a driving force in the "Seattle Citizen's Committee's" plan to shut down World Trade Organization talks. DAN's ethos is nonviolent protest, and the group denounces the cops for sparking Seattle's street infernos and curfew alerts. DAN *détourns* high-tech media and works it for its own ends, coordinating on the Internet, initiating guerrilla action, radicalizing fellow-traveling affinity groups into a singular contesting force. Other member groups, like the Ruckus Society, affirm a politics of pleasure, having fun while getting serious, performing street theater and musical happenings, dancing soirees, and holding educational seminars.

Many cities across the globe have also been disrupted and reappropriated by another dynamic spontaneous presence: Reclaim the Streets (RTS). In recent years, RTS demos have shut down streets in Manhattan (at Astor Place, in the East Village, and around Times Square); in Sydney; in north, south, and central London; in Helsinki; and in Prague and other European capitals. In the middle of major traffic thoroughfares, crowds have danced and shouted and partied—revolutionaries, students, workers, activists, madmen, and malcontents. In their "Festivals of Love and Life," they've brought cars to a standstill and demanded pedestrians' and bikers' right to the city. In New York, they rallied against ex-mayor Rudy Giuliani's "quality of life" campaigns against the homeless, the sidewalk vendors, and the poor. In Seattle, under the noses of neoliberal bigwigs, RTS clasped hands with Global Exchange to embarrass the hell out of politicians and business honchos plotting to carve up the world into profit centers. RTS has rediscovered a "new romantic" Lefebvrian oomph, "transforming stretches of asphalt into a place where people can gather without cars, without shopping malls, without permission from the state, to develop the seeds of the future in the present society." So said one RTS poster I saw not so long ago on an East Village wall.

RTS began in London in 1991 when people banded together to contest the Conservative government's large-scale highway construction program, a hair-brained policy destined to slice huge swaths through verdant countryside and vibrant cityscape. Before long, a concerted antiroads campaign surfaced over the fate of Twyford Down, near Winchester, where rolling pastures and ancient walkways stood in the path of the proposed (and subsequently completed) M3 extension between Southampton and

London. For several years, Twyford Down was a war zone and a radical cause célèbre. Ironically, antiroad mobilizations and RTS activism grew in the face of Tory legislation explicitly engineered to stamp it out: the 1994 Criminal Justice Act (CJA), which tried to outlaw any public gathering or street "disorder" involving twenty or more people. In simple terms, anything that didn't figure on then prime minister John Major's "democratic" agenda, like genuine free speech and collective protest, could henceforth be rendered illegal. (The CJA still persists in Blair's Britain.) After its inception, the CJA duly fanned the flames of its "other," being increasingly imposed on increasing numbers of public gatherings condemning the CJA.

RTS/London emerged within this adversarial atmosphere, staging its first "street party" at busy Camden High Street in north London in 1997. The following year, just down the street, it sealed off an even busier artery adjacent to King's Cross Station: dancers motioned to drumbeats, and hoards of different sorts of people hung out and reclaimed for pedestrians a big stretch of Britain's capital. By that time, the RTS concept had a distinctive West Coast drawl, touching down in Berkeley, where RTS/Bay Area liberated Telegraph Avenue for a while. Then, responding to Giuliani street cleanup vendettas, RTS/New York came of age in the Big Apple, begetting "great feasts of public space." Suddenly, protest became imaginative and fun again, veritable be-ins and "carnivals of freaks," contesting zero tolerance policing, privatization, and sanitization of city life and appealing instead for real human rights, for real public space. Central to RTS's modus operandi is play and festival, as it is for a lot of the antiglobalization movement.

Such prankster politics enacts lampoon, pulls tongues and raises the finger, and voices satire at a rather sober and stern enemy. Turning people on has often meant turning them off party-political smokescreens. They know the revolution will never be televised. Meanwhile, protagonists have recognized a common fate and common foe as they've explored a common opportunity. Lefebvre never saw any of these battles and ransackings, but one wonders what he would have made of them. This time around the hairstyles and fashions of the protesters are different, and they speak in a different tongue and jostle a new-fangled enemy. But they remain Lefebvrian at heart: spontaneous yet smart, politically savvy as well as theoretically astute, Rabelaisian revelers reconstructed.

4

URBANITY

> The antagonism between the town and country can only exist within the framework of private property. It is the most crass expression of the subjection of the individual under the division of labor, under a definite activity forced upon him—a subjection which makes one man into a restricted town-animal, the other into a restricted country-animal, and daily creates anew the conflict between their interests.
>
> —**Karl Marx and Frederick Engels,** *The German Ideology*

There's little doubt that the encounter with Guy Debord and the Situationists piqued Henri Lefebvre's interest in things urban. Hitherto, the peasantry and the countryside had captured his imagination; hitherto, his Marxist social and political theory had critiqued fascism and pilloried state socialism, burrowed into alienation, and affirmed everyday life; hitherto, he'd posed new utopian questions and proposed old romantic solutions, indicted capitalist modernity in the name of a new, more spontaneous modernity—one with medieval roots. Now, Lefebvre's concrete abstraction became the modern city itself, the testing ground for new Marxist thinking and utopian radical praxis. "The urban" became at once the dread zone and the nemesis of capitalist modernity, the cradle of unprecedented commodification as well as the incubator for new experimental lived moments. Curiously, raw data had been in front of Lefebvre's nose for a long while; the Situationists merely helped him correct his myopia, for he'd seen it all coming in his own daily life in Navarrenx and nearby in a town called Mourenx. Henceforth, in the "Seventh Prelude" of *Introduction to Modernity,* in "Notes on the New Town," he began to tell us what he saw, what was wrong, and what might be right.

* * *

"Whenever I set foot in Mourenx," Lefebvre says, "I am filled with dread." Mourenx is a prototypical species, a French New Town, which, like other New Towns then sprouting up on the European (and American) landscape, "has a lot going for it."[1] He thinks,

> The overall plan has a certain attractiveness: the lines of the tower blocks alternate horizontals and verticals. ... The blocks of flats look well planned and properly built; we know that they are very inexpensive, and offer their residents bathrooms or showers, drying rooms, well-lit accommodation where they can sit with their radios and television sets and contemplate the world from the comfort of their own homes. ... Over here, state capitalism does things rather well. Our technicists and technocrats have their hearts in the right place, even if it is what they have in their minds which is given priority. It is difficult to see where or how state socialism could do any differently or any better.[2]

Still, every time he sees these Le Corbusian "machines for living in," he's terrified, adamant that such a new mode of life is Cartesian through and through, compartmentalizing different facets of human activity, zoning things here and there, creating functional spaces and atomized people who are turned inward, away from one another, even though they're often piled on top of one another.

It is in Mourenx, Lefebvre says (p. 119), where "modernity opens its pages to me." There, rational knowledge, technological ingenuity, and a Logos big-brain fix to pressing human needs equates to *separation*—of people and activity—all done in the name of efficiency and profitability. Lefebvre, as ever, is less interested in economic machinations than with metaphysical misgivings. He invokes the young Marx and a left-wing Hegel, both of whom strove to reconcile the Cartesian partitioning of mind and matter, of subject and object, rather than reify it in physical space. For Lefebvre, every New Town, every new suburb—every Levittown, Middletown, or Our Town emerging out of the rubble—has hacked up space and simplified life, decanted people, and flattened experience. At the same time, separation means separation within the self, a partitioning of consciousness, an inability to connect organically with what's around you, to think the whole, to understand the totality of your life—or to not want to understand it anymore. As Lefebvre sees it, planners and technocrats, in cahoots with bankers, constructors, and realtors, have somehow become new "Grand Inquisitors," profiting financially and politically from modernization, promising people bread and security as long as they can stealthily control their freedom.

The accusation redoubles Lefebvre's commitment to Marxist humanism, only now this commitment has a territorial embodiment, is conceived as a *spatialized* Marxist humanism. Now, a more wholesome personhood is predicated on a more wholesome organization of urban and rural space. In the course of its long history, from the ancient Greeks to the Middle Ages, the city, Lefebvre points out, was once an inspiring organic unity, intimately bonded with the countryside; the two realms coexisted in a delicate but real symbiosis. Now, this symbiosis, this organic unity has been undone, dismembered, dislocated.[3] Both the city and the countryside are victims of the inexorable drive to accumulate capital, a drive orchestrated by assorted agents and agencies of the capitalist state. Everyday life had become at once colonized, fragmented, and politicized. Once, in Greek times, with its dynamic public-square *agoras,* the *polis* epitomized the very essence of civil society in harmony with the state. "The state coincided with the city and civil society," Lefebvre says, "to form a polycentric whole, and private life was subservient to it."[4] It wasn't until the "modern world," as the young Marx highlighted, that the *abstraction of the state* and the *abstraction of private life* were born.[5] Marx used the term *modern* to periodize the rise of the bourgeoisie, the development of industrial growth, and the "real subsumption" of modern capitalist production. Between 1840 and 1845, Marx pinpointed, in effect, the birth of modern modernity. The type of the state Marx defined, Lefebvre explains, "is one which separates everyday life (private life) from social life and political life. … As a result, private life and the state—that is, political life—fall simultaneously into identical but conflicting abstractions" (p. 170).

For Lefebvre, Mourenx demonstrates how fragmentation and conflicting abstractions materialize themselves in bricks and mortar—and in plastic. In modern everyday life, streets and highways are more and more necessary to physically connect people, "but their incessant unchanging, ever-repeated traffic is turning [human space] into wastelands" (p. 121). Everything seems topsy-turvy: "Retail is becoming more important than production, exchange more important than activity, intermediaries more important than makers, means more important than ends" (p. 121). Strangely, there aren't many traffic lights in Mourenx, even though the whole town seems "nothing but traffic lights" (p. 119). Mourenx's physiognomy is left naked, robbed of meaning, "totally legible." Here, as elsewhere, a "stripping process" has been accomplished. "Every object," Lefebvre says, "indicates what its function is, signifying it, proclaiming it to the neighborhood. It repeats itself endlessly. When objects are reduced to the basic level of a signifier they become indistinguishable from things *per se*. ... Surprise? Possibilities?" (p. 119). What, he asks, are we on the threshold of? "Are we entering a Brave New World of joy or a world of irredeemable boredom? As yet I cannot give an answer" (p. 119). At any rate, one conclusion is immediately evident: Mourenx's world—the world of the high-rise New Town and the low-rise suburb—expresses an ordered, enclosed, and *finished* world, a world in which there's nothing left to do and nobody to pull tongues at, no romance around any corner. What's there is simply there.

Enter, by comparison, Navarrenx, barely ten miles from Mourenx in one sense, yet light-years away in another. In the fourteenth century, it too was a New Town, built to a fairly regular ground plan near the Pyrenean River Oloron and rebuilt two centuries later in an even more geometric design, ringed with Italianate ramparts. "I know every stone of Navarrenx," Lefebvre tells readers (p. 116); in each stone, he reads, like a botanist reads the age of a tree by the rings on its trunk, centuries of history, histories entombed in space. But past voices still speak out volubly to the living. Lefebvre likens Navarrenx's subtle and instructive development—the symbiosis between its physical and social growth—to that of a seashell. A seashell is the product of a living creature that's slowly "secreted a structure." Separate the creature from the form it's given itself—relative to the laws of its species—then you're left with something soft, slimy, and shapeless. Thus, the relationship between the animal and the shell is vital—quite literally—for understanding both the shell and the animal. Navarrenx's shell, Lefebvre says, embodies the forms and actions of a thousand-year-old community, "shaping its shell, building and rebuilding it, modifying it again and again and again according to its needs" (p. 116). History and civilization is a sort of seashell, he reckons. Look closely at every medieval town, every little house, every winding cobbled street, every courtyard or square, every passageway or back alley and you'll see the mucous trace of this animal who "transforms the chalk in the soil around it into something delicate and structured."

Not every city or town, of course, has the luxury of either a deep past or a rich culture, especially in the North American New World. And yet, it is precisely the link between the animal and its shell that Lefebvre forces us to consider, the form of the dilemma as much as the content itself. The keyword here is *unity*. Lefebvre is mesmerized by Navarrenx's organic unity, by its intimacy, by its style and function—by its "charm," he jibes. Everything about it has unity, is a seamless whole; everything relates, leads us into another space with another function, and then onto an another, and another, with a completely different function. "There is no clear-cut difference," he writes, "yet no confusion exists between the countryside, the streets and the houses; you walk from the fields into the heart of the town and the buildings, through an uninterrupted chain of trees, gardens, gateways, courtyards and animals" (p. 117). What's more, streets aren't wastelands, nor are they simply where people go from A to B. They are places "to stroll, to chinwag [*bavarder*], to be alive in." With a tonality reminiscent of Jane Jacobs (whose *Death and Life of the Great American Cities* was published exactly the same moment Lefebvre penned this critique), "nothing can happen in the street without it being noticed from inside the houses, and to sit watching at the window is a legitimate pleasure. ... The street is something integrated" (p. 117).

Sadly, Navarrenx, like many small towns and villages the world over, has been dying for a while now, victim of changes in industry and agriculture, trampled on by the march of "progress";

"the expiring seashell," Lefebvre laments, "lies shattered and open to the skies" (pp. 117–18). Likewise it has gotten more boring as time has passed. Navarrenx's market day is tiny compared with those of yesteryear; surviving storekeepers are little more than managers now; narrow streets are gridlocked each day with cars and trucks. Nevertheless, its boredom is more complacent, softer, and cozier, more comforting and carefree than Mourenx; it's the boredom of a lazy summer Sunday afternoon or a long winter night in front of a roaring open fire. Mourenx's boredom, conversely, "is pregnant with desires, frustrated frenzies, unrealized possibilities. A magnificent life is waiting just around the corner, and far, far away. It is waiting like the cake when there's butter, milk, flour and sugar." In Mourenx, "man's magnificent power over nature has left him alone to himself" (p. 124). This is a thoroughly modern boredom, one affecting heavily the youth, those without a future, and women, who always, Lefebvre says, bear the brunt of an isolated and dismembered everyday life.

Lefebvre can't hide his admiration of old medieval towns. And who can blame him, given the ugly giant sprawls we today call cities? By choice or default, large numbers of us have lives that open out onto vast voids of desolation and nothingness. But Lefebvre's alternative warrants caution. At times, his fondness smacks of *gemeinschaft* nostalgia, a romantic yearning for paradise lost, for a bygone age when everything was unified and whole, artisanal and authentic. He knows he's treading through Proudhonian minefields.[6] Yet it soon becomes evident he has something else in mind. The metaphor of the seashell is crucial. With it, Lefebvre wants to emphasize the relationship between an animal (i.e., human beings) and its habitat (i.e., our cities), specifically how the habitat should be flexible enough to permit free growth of the animal, responsive enough to "the laws of its species."[7] Growth of an animal, he says, follows a certain functioning order. And in the case of human beings, we produce our lives *knowingly* and self-consciously, making us special, gifted animals, different from seashells—smarter, right? We're beings who *should* know what's good for us. Marx tried to redouble the point long ago: "A spider conducts operations which resemble those of the weaver, and a bee would put many a human architect to shame by the construction of its honeycomb cells. But what distinguishes the worst architect from the best of bees is that the architect builds the cell in his mind before he constructs it in wax."[8]

Our uniqueness means we have two distinctive ways of creating and producing—of secreting our structure. Hitherto, Lefebvre says, they've rarely coincided: a spontaneous-organic method and an abstract, a priori approach of planning for rainy days ahead. So the dilemma: How to cultivate spontaneity? How to create a spontaneous life-form out of an abstraction? How can we create an urban culture based around both lived practice and conceived premeditation, learning from the past while experimenting with the future? Before technology penetrated everyday life, before capitalist industrialization used it to begat a bastardized form of urbanism, everyday life "was alive. The slimy creature secreted a beautiful shell" (p. 123). "It is impossible," he reflects (p. 122), when stood atop a small hillock above Mourenx, surveying the modern works down below (like Faust in Part II of Goethe's great fable), "looking ridiculous" as only a Left intellectual can, "not to be reminded of what Marx wrote [in *The German Ideology*] when he was still a young man: 'Big industry … took from the division of labor the last semblance of its natural character. It destroyed natural growth in general … and resolved all natural relationships into money relationships. In place of naturally grown towns it created the modern, large industrial cities which have sprung up overnight.' " Can spontaneity ever be revitalized in Mourenx? Lefebvre asks. Can a community be created—can it create itself? How can we humans, in a new millennium, having gone to the moon and cloned ourselves, reconcile organicism with prefigurative ideals?

Is the city a technical object or an aesthetic moment, an œuvre or a product?

* * *

Lefebvre began to confront these questions head on—with typical "cavalier intention"—in a provocative text, *The Right to the City* (1968), a series of exploratory essays, drafted during the 1960s (and updated and upgraded in 1972). Here the aging Rabelaisian Marxist unveils for the first time, as a coherent whole, his analysis on an emergent urban society.[9] A "double process" (p. 70) is before us, he says near the beginning: "*industrialization* and *urbanization,* growth and development, economic production and social life." An inseparable and inexorable unity has been born, a terrible Janus-faced beauty, coexisting in Manichean disunity, pitting industrial reality against urban reality, a mode of production against its built form: a rabid animal is set to burst out of its beat-up shell.

Industrialization, Lefebvre reminds us, produces commodities at the same time as it proletarianizes people, creates wealth while it needs to reproduce its workforce, somewhere. The process spawns fields and factories, haciendas and housing estates, bosses and managers, bank districts and financial centers, research complexes and political power hubs. All of which prizes open, and hacks up, urban space itself, transforming the countryside to boot, reforging everything and everywhere on the anvil of capital accumulation. To "manage" an unmanageable contradiction, a new crew of frauds enters the fray: planners and politicians, technocrats and taskmasters, who speak a new "discourse," Lefebvre says, replete with a new *ideology:* that of *urbanism.* Orchestrated by the state, the urban question henceforth becomes a *political* question; class issues are now explicitly urban issues, struggles around territoriality, out in the open.

To some extent, that's the good news. As for the bad news, the urban fabric [*tissu urbain*] has been mortally wounded, sliced into like live flesh, leaving amputated body parts and a whole lot of blood. "Populations are heaped together," Lefebvre notes, "reaching worrying densities. At the same time, old urban cores are deteriorating or shattering. People are displaced to far-off residential or productive peripheries. Offices replace housing in urban centers. Sometimes (in the United States) centers are abandoned to the 'poor' and become ghettoes of the disenfranchised. Other times, the most affluent people retain their stake at the heart of the city" (p. 71). The city has become either recentered or decentered, asphyxiated or hollowed out, a showcase or a no place. Consequently, it hasn't just lost a sense of cohesion and definition; its dwellers have lost a sense of *creative and collective purpose*.

Cities are little more than places where people earn money, speculate on money, or merely live. What should be stunning projects that people *inhabit* have become dismal *habitats,* seats of decivilization. Lefebvre uses here the term *inhabiting* to stamp a richer gloss on city life, evoking urban living *as becoming,* as growing, as something dynamic and progressive. Being in a city, he stresses, is a lot more than just being there. The nod, duly acknowledged, is made to Martin Heidegger (1889–1976): "To inhabit," Lefebvre explains, "meant to take part in a social life, a community, village or city. Urban life possessed, amongst other qualities, this attribute. It bestowed dwelling, it allowed townspeople-citizens to inhabit. It is thus that 'mortals inhabit while they save the earth, while they wait for gods ... while they conduct their own being in preservation and use.' Thus speaks the poet and philosopher Heidegger of the concept to inhabit" (p. 76).

But Lefebvre loosens the deep ontological moorings of the German philosopher's notion of "place as the unique dwelling of being" and beds the concept down in political and historical reality.[10] As such, a loss of inhabiting is a political, social, and aesthetic loss. Downgrading *inhabit,* reducing it to a mere *habitat,* signifies a loss of the city as *œuvre,* a loss of *integration* and *participation* in urban life. Indeed, it is to denigrate one of humanity's great works of art—not one hanging on a museum wall but a canvas smack in front of our noses, wherein we ourselves are would-be artists, would-be architects.

In those sections on inhabiting, and on the city as œuvre, Lefebvre writes beautifully, and inspiringly, about the urban, invoking the power of the city, the promise of the city, more as an artist intent on pleasure than as a sociologist intent on measure. Always his target is a bigger virtue, a deeper understanding of human reality; always he blurs together past, present, and future, conceiving the city as a historical as well as a *virtual* object, something that's simultaneously disappeared and yet to appear. Conjecture pops up as quickly as fact—a trait destined to irk, or befuddle, traditional social scientists, those motivated by *is* rather than *ought.*

That the city is "an exquisite *œuvre* of praxis and civilization" (p. 126) makes it very different from any other product. "The *œuvre,*" Lefebvre insists, "is use value and the product is exchange value. The eminent use of the city, that is, of its streets and squares, buildings and monuments, is *la fête* (which consumes unproductively, without any other advantage than pleasure and prestige)" (p. 66). And this unproductive pleasure was a free-for-all, not a perk for the privileged. Needless to say, cities throughout time have been seats of commerce, places where goods and services are peddled, spaces animated by trade and rendered cosmopolitan by markets. Middle Age merchants, Lefebvre confirms, "acted to promote exchange and generalize it, extending the domain of exchange values; yet for them the city was much more than an exchange value" (p. 101). For sure, it's only a relatively recent phenomenon that cities *themselves* have become exchange values, lucre *in situ*, jostling with other exchange values (cities) nearby, competing with their neighbors to hustle some action—a new office tower here, a new mall there, rich *flâneurs* downtown, affluent residents uptown.

Industrialization has commodified the city, set in motion the decentering the city, created cleavages at work and in everyday life: "Expelled from the city," Lefebvre writes, "the proletariat will lose its sense of *œuvre*. Dispensable from their peripheral enclaves for dispersed enterprises, the proletariat lets its own conscious creative capacity dim. Urban consciousness vanishes" (p. 77). "Only now," reckoned Lefebvre in the 1960s, "are we beginning to grasp the *specificity* of the city" (p. 100), a product of society and of social relations yet a special feature within those relations. Indeed, urbanization now *reacts back* on society, for better or for worse, and has run ahead of industrialization itself. It's only now, he adds, using his own emphases, that the "foremost theoretical problem can be formulated": "For the *working class,* victim of segregation and expelled from the traditional city, deprived of a present and possible urban life, a practical problem poses itself, a *political* one, even if it hasn't been posed politically, and even if until now the housing question … has masked the problematic of the city and the *urban*" (p. 100).

The latter allusion, of course, is to Engels's famous pamphlet *The Housing Question* (1872), in which Marx's faithful collaborator denounced those petty-bourgeois reformists who wanted to resolve squalid worker housing conditions without resolving the squalid social relations underwriting them. While Lefebvre, a fellow Marxist, concurs with Engels's analysis and critique, as well as with his *political* reasoning, he can't, circa late twentieth century, adhere to Engels's solution:

> The giant metropolis will disappear. It should disappear. Engels possessed this idea in his youth and never let it go. In *The Housing Question,* he'd already anticipated, "supposing the abolition of the capitalist mode of production," an equal

> as possible repatriation of the population over the entire land. His solution to the urban question precludes the big modern city. Engels doesn't seem to wonder if this dispersion of the city throughout the surrounding countryside, under the form of little communities, doesn't risk dissolving "urbanity" itself, of ruralizing urban reality.[11]

In truth, "there can't be any return to the traditional city," Lefebvre rejoins (p. 148), notwithstanding his affection for Navarrenx, notwithstanding his admiration for Engels, just as there can't be any "headlong flight towards a colossal and shapeless megalopolis." What we must do, he says, is "reach out and steer ourselves towards a new humanism, a new praxis, another man, somebody of urban society" (p. 150). This new humanism will be founded on a new right, the right to an œuvre, *the right to the city,* which will emerge "like a cry and demand," like a militant call to arms. This isn't any pseudo right, Lefebvre assures us, no simple visiting right, a tourist trip down memory lane, gawking at a gentrified old town; neither is it enjoying for the day a city you've been displaced from. This right "can only be formulated," he says, "as a transformed and renewed right to urban life" (p. 158), a right to renewed centrality. There can be no city without centrality, no urbanity, he believes, without a dynamic core, without a vibrant, open public forum, full of lived moments and "enchanting" encounters, disengaged from exchange value. "It doesn't matter," he says, "whether the urban fabric encroaches on the countryside nor what survives of peasant life, so long as the 'urban,' place of encounter, priority of use value, inscription in space of a time promoted to the rank of a supreme resource amongst all resources, finds its morphological base, its practical-material realization" (p. 158).[12]

Asserting his hard-core Marxist credentials, at the centenary of Marx's *Capital* (1967), only a united working class, concludes Lefebvre in a series of "Theses on the City," has the power and wherewithal to reappropriate the city, to be the "bearer" of a new "virtual action" (p. 179), a new urban praxis in "the general interests of civilization." Yet the working class hasn't hitherto discovered a spontaneous sense of the city as œuvre; its conscious has dimmed, has almost disappeared with artisan and craft workers. From where can it summon up this collective spontaneous sense, reclaiming its œuvre *really* not virtually? How can the working class become the bearer of this higher consciousness, using "its productive intelligence and its practical dialectical reason"? Can we demand of the working class the possible and impossible, a concrete as well as an experimental utopia? So many questions: a typical Lefebvrian mode of argumentation. As for responses, was it not Marx, he queries, who once said humanity poses only problems it knows it can resolve? Sometimes, the solutions are already there, not faraway, waiting for questions yet to be asked.

* * *

Much of Lefebvre's urban theory was based on firsthand experience, gleaned from prodigious travel and lecturing schedules. After his "retirement" from Nanterre in 1973, this was tantamount to a round-the-world tour, an epic global *dérive*. (The Guterman Archive preserves the glossy, bright-colored postcards from Lefebvre's roving travel chest.) While Lefebvre loved to think big and make grand, sweeping abstractions, he was intimately acquainted with the world's great cities. He never tired of discovering new places; his geographical curiosity abated not, even as an old man. In younger days, he'd sojourned in London and Amsterdam, in Barcelona and Berlin; with the party, he'd toured the old Eastern Bloc, explored Poland and Bulgaria and Yugoslavia, Rumania, and Hungary, knew its capitals Warsaw and Sofia and Belgrade, Bucharest, and Budapest. (The heretic humanist never made it the USSR, though, never visited Moscow.

"I would've liked to have taught in Moscow," he admitted in 1988. But each time he tried, "they always denied it me.")[13] He journeyed to Italy, adored Venice and Florence; he traveled throughout North America and South America, went to New York (with Norbert Guterman) and Los Angeles, to Montreal and Toronto; he lectured in Mexico City and Santiago; in San Paulo, Rio, and Brasilia; in Caracas and Buenos Aires. In Africa, he knew Algiers and Tunis, Casablanca and Dakar. He toured around Iran and China, discovered Tehran and Shanghai and Beijing; went to Japan and Tokyo and onward on to Australia and Sydney.

In 1983 to 1984, at the invitation of literary critic Fredric Jameson, Lefebvre spent a semester in the History of Consciousness Program at the University of California, Santa Cruz, and deepened his fascination—and disgust—with West Coast–style urbanization. "It's extremely difficult to give an answer to the question of which city one likes and dislikes," he once owned up, "for detestable cities are intriguing. Take Los Angeles. For a European it's appalling and unlivable. You can't get around without a car and you pay exorbitant sums to park it. ... What fascinates and disgusts me are the streets of luxury shops with superb windows but which you can't enter into. ... These streets are empty. And not far from there, you have a street, a neighborhood, where 200,000 Salvadorian immigrants are exploited to death in cellars and lofts." Yet there is "singing and dancing," he says, "something stupendous and fascinating. You are and yet you're not in a city, stretching for 150 kilometers, with twelve million inhabitants. Such wealth! Such poverty!" At the same time, "you feel that the Hispanics have a counterculture, and they make the society, the music, painting (the murals they've created are beautiful)."[14] Lefebvre took numerous trips to Los Angeles. One time, he and Jameson (who was then working on his Bonaventure Hotel/Postmodernism article), together with UCLA geographer–planner Ed Soja, did a downtown tour. "He was fascinated," recalled Soja, "particularly by the

Estrada Courts public housing project, where nearly all the walls are covered with murals, the most notable being a stark picture of Che with the admonition 'We are not a Minority!' "[15]

Cities attaining the heady status of œuvres nonetheless remained dearest to Lefebvre's heart. Venice is adored, a city reshaped by time and literally receding into the sea, yet living on as a great work of art, as an architectural and monumental unity, with its misty, haunting melancholy, sound tracked by Mahler's 5th Symphony; it's a city, says Lefebvre, at once "unique, original and primordial," despite the tourists, despite its "spectacularization."[16] Every bit of Venice "is part of a great hymn to diversity in pleasure and inventiveness in celebration, revelry and sumptuous ritual."[17] Is Venice "not a theatrical city, not to say a theater-city—where actors and the audience are the same in the multiplicity of their roles and relations? Accordingly, one can imagine the Venice of Casanova, and Visconti's *Senso* [and *Death in Venice*], as the Venice of today."[18]

Lefebvre's favorite city, however, is Florence, beside the Arno, a "symbolic flower," immortalized in *Lorenzaccio* (1834) by one of his heroes, Alfred de Musset. ("The banks of the Arno are full of so many goodbyes," said Musset.) "Florence has ceased recently to be a mummified city, a museum city," Lefebvre said in 1980, "and has found again an activity, thanks to small industries on its periphery."[19] "So what I like is Los Angeles for the fascination, Florence for the pleasure and Paris to live in."[20]

Even as an octogenarian, Lefebvre continued to probe the city, reached out into seemingly uncharted theoretical territory. His urban fascination never relented, even though he'd seen it all, perhaps many times over. He marveled at the everyday rhythms that ripple and syncopate urban life and, as such, coined a new theoretical practice: *rhythmanalysis,* the eponymous title of his final book, written with wife Catherine Regulier. All of which heralded, in his own words, "nothing less than a new science, a new field of knowledge: the analysis of rhythms, with practical consequences."[21] The idea of a rhythm was deliberately provocative, an assault on those who reify the city as a thing, who document only what they see rather than what they feel or hear. Movement and process, along with frequency and melody, now became Lefebvre's muse. Here the old man listened to urban murmurs much the same way he tuned into Schumann's *Carnaval* or Beethoven's 9th. Secret rhythms, buried in the city's subconscious, are unearthed by Lefebvre, as are public rhythms that chime in the agora, get instantiated by festivals and mass celebration; fictional rhythms ring out, too, those invented in Lefebvre's own aging imagination, as he loses himself out of his own apartment window, staring down on everyday Paris. Rhythmanalysis signals an ancient scholar's farewell, his last gasp, an indulgence we can forgive, even when we know very little adds up or extends what he's told us already. Rhythmanalysis was Lefebvre's personal right to city, a right he perhaps should never have shared.

* * *

In the fall of 2004, the French newspaper *Libération* (September 16, 2004) headlined the findings of the United Nations–Habitat's "World Urban Forum," held in Barcelona. "THE DAMNED OF THE CITY" ran the bleak front-page leader. In 2020, two billion people are projected to inhabit assorted shantytowns, *favelas* and *bidonvilles,* and the majority of our megacities will burgeon into spaces of the poor—gigantic, sprawling neighborhoods of cardboard and tin, of prefabricated materials destined to be washed away in the next mudslide. By 2015, twenty-three cities will have populations in excess of 10 million, with Tokyo (26.4 million), Bombay (26.13 million), Lagos (23 million), Dacca (21.1 million), São Paulo (20.44 million), Mexico City (19.2 million), and Karachi (19.2 million) topping the premier league. Of those twenty-three cities, nineteen will be in areas of the planet deemed either "underdeveloped" or "developing." By 2050, forty-nine of the world's poorest countries will have tripled their populations, resulting in an exponential growth in urban slums.

"It's a planetary revolution," said *Libération*'s editorial "Exodus" (p. 2), "destructive like all revolutions, and nothing is controlling its destabilizing effects. An exodus that forces millions of human beings to quit rural zones, where humanity has lived since prehistory, towards the megalopoles more and more monstrous and chaotic. … At the start of the 21st century, an immense migration, fuelled by the fascination of the city's bright lights and the hope of escaping the stupefying misery of the countryside, accelerates and extends urbanization to the furthest reaches of the planet." In two years time, "for the first time in our history," the report noted, "the majority of humanity will dwell in cities. … Inequality and injustice, misery and violence, criminality and corruption, are the price of this mutation, which economic globalization amplifies."

It was seemingly for good reason, with thirty-five years hindsight, that Lefebvre kicked off his greatest urban text, *The Urban Revolution,* with a chapter called "From the City to Urban Society." "We will depart from a hypothesis," he began, which needs supporting by argument and evidence: "society has been completely urbanized." Back then, Lefebvre thought we should speak *not of cities* but of *urban society*—a "virtual reality," he wrote in 1970, yet "tomorrow real."[22] That tomorrow is already our today. A society born of industrialization has indeed succeeded industrialization, has at once realized and surpassed it, has made it somehow "postindustrial," a point of arrival as well as a point of departure. The United Nation's Habitat program is set to look down the abyss, poised to address this "urban gangrene." Already they know a thing or two: "The promotion of participatory and inclusive styles of local governance," United Nation–Habitat's

"Executive Summary" advised, with an uncanny Lefebvrian timbre, "has proved to be an effective means of … overcoming poverty." "There are a number of methods," the summary stressed, "developed and adopted that maximize benefit of inclusive governance and offset the opportunity costs to the poor. It was noted that inclusion is guaranteed when every urban citizen has a 'Right to the City.' "[23]

5

Urban Revolution

> It is in the countryside that seditious thought ferments, but it is in the city that such thought erupts. Liberty likes extreme crowds or absolute solitude.
>
> —**Louis Gauny,** *Le philosophe plébéien*

The Urban Revolution did indeed represent an arrival as well as a point of departure, for both Henri Lefebvre and the world. Here was a book rooted and incubated in the tumult of 1968 yet anticipated much more a new era ahead, a post-1968 age, replete with its cynicism and promises, its possibilities and impossibilities. *The Urban Revolution* marked a new beginning, the dawn of a thoroughly *urbanized society;* "the urban" entered the fray like the Nietzschean "death of God" or the Marxian "loss of halo": all hitherto accepted values and morals had been drowned in economic and political ecstasies, in postwar exigencies threatening everybody. By 1970, Lefebvre recognized the hopes of those street-fighting years were dashed and knew a sober reconceptualization was warranted. At the same time, he can't quite give up the ghost. The moment engendered new opportunities, fresh chances to revalue all values, to invent a "new humanism." Such is Lefebvre's wish-image of a future awaiting its metropolitan birth pangs. The urban became his *metaphilosophical* stomping ground, the contorted arena of new contestation and reinvented Marxist practice.

The notion of "revolution," of course, has a sixties swing about it, and Lefebvre knows his explosive little title will stir the left as much as the right. In fact, his book sought to lodge itself within each flank, just as it intended to detonate both. "The words 'urban revolution,' " he writes, playfully, "don't in themselves denote actions that are violent. Nor do they exclude them."[1] The revolution Lefebvre simultaneously comprehends and aims to incite is a *process* as well as *praxis,* a theoretical and a practical problematic. What he wants to comprehend is a revolution that his Marxist bedfellow Antonio Gramsci might have labeled "passive"—a revolt instigated from above, a counterrevolution. What he wants to incite is an urban revolution more akin to the Paris Commune, what Gramsci might have called a "war of position," a popular and historical assault from below. The *process* Lefebvre reveals comprised immanent contradictions festering within global capitalism, those about to blow on the cusp of Keynesian collapse. Meanwhile, this *praxis* had its own ideological thrust and institutional base—both free-market and left-wing-technocratic, which, in the decades to follow, would congeal into a single neoliberal orthodoxy. Thus, Lefebvre's key urban text has a prescient subtext: it identifies the structural collapse of industrialism and state managerialism wherein urban revolution symbolizes a "postindustrial" revolution, a society no longer organized by planners but speculated on by entrepreneurs, a society we know to be our own.

* * *

We must grasp the whole and take this new reality by the root, Lefebvre says. Neither analytical fragmentation nor disciplinary compartmentalization will do. A new order is evident, which knows no restrictions and breaks through all frontiers, overflowing everywhere, seeping out across the world and into everyday life. Critical theory and left politics must respond in kind, thinking big and aiming high, or else it will aim too low and give up on getting even that far. We need to be "revolutionary," Lefebvre insists, because what we have before us is revolutionary. Like Marx's inverting Hegel to discover the "rational kernel" within the "mystical shell," Lefebvre stands mainstream economic and sociological wisdom on its head: "we can consider industrialization as a stage of urbanization, as a moment, an intermediary, an instrument. In the double process (industrialization-urbanization), after a certain period the latter term becomes dominant, taking over from the former" (p. 185; p. 139). As the mainstay of a capitalist economy, urbanization has supplanted industrialization, he reckons. The capitalist epoch reigns because it now orchestrates and manufactures a very special commodity, an abundant source of surplus value as well as massive means of production, a launch pad as well as a rocket in a stratospheric global market: urban space itself.

We must no longer talk of *cities* as such, Lefebvre urges; all that is old hat. Rather, we must speak of *urban society,* of a society born of industrialization, a society that shattered the internal intimacy of the traditional city, that gave rise to the giant industrial city Engels documented, yet has itself been superseded, been killed off by its own progeny. Industrialization, in a word, has negated itself, bitten off its own tail, advanced quantitatively to such a point that qualitatively it has bequeathed something new, something dialectically novel, something economically and politically *necessary.* "Economic growth and industrialization," Lefebvre writes, "have become at once causes and crucial reasons, extending their effects to entire territories, regions, nations and continents." Absorbed and obliterated by vaster units, rural places have become an integral part of industrial production, swallowed up by an "urban fabric" continually extending its borders, ceaselessly corroding the residue of agrarian life, gobbling up everything and everywhere that will increase surplus value and accumulate capital. "This term, 'urban fabric,' " he qualifies, "doesn't narrowly define the built environment of cities, but all manifestations of the dominance of the city over the countryside. In this sense, a vacation home, a highway and a rural supermarket are all part of the urban tissue" (p. 10; pp. 3–4).

What's fascinating here is how *The Urban Revolution* appeared only a year before U.S. president Richard Nixon devalued the dollar, wrenched it from its gold standard mooring to herald the United States's unilateral abandonment of the 1944 Bretton Woods agreement. Gone, almost overnight, was the system of financial and economic regulation that spearheaded a quarter of a century of capitalist expansion. As the U.S. economy bore the brunt of a costly and nonsensical war in Vietnam, 1971 ushered in an American balance of trade deficit. Nixon knew fixed exchange rates couldn't be sustained, not without overvaluing the dollar, not without losing competitive ground. So he let the dollar drift, devalued it, and loosened Bretton Woods's grip. World currency hereafter oscillated; capital could now more easily slush back and forth across national frontiers. A deregulated, unstable capitalism became rampant, without restraint, and Lefebvre sensed its coming, saw how it facilitated what he'd call the "secondary circuit of capital," a siphoning off of loose money set on speculation in

“The Head of Don Quixote”: Henri Lefebvre, 1978.

"Greetings from Navarrenx!"

Brueghel, Pieter the Elder, *Fight Between Carnival and Lent*, 1559. Kunsthistorisches Museum, Vienna, Austria. Erich Lessing/Art Resource, NY.

The Explosion: Paris, May 1968. © Bettmann/CORBIS

The Pompidou Center: from Lefebvre's front door, rue Rambuteau, Paris. Photo by Andy Merrifield.

Chez Lefebvre, 30, rue Rambuteau, Paris. Photo by Andy Merrifield.

PARIS le 18 Octobre

Cher Norbert,

Cinq mois déjà que j'ai quitté Brewster.Quatre mois et demi au moins que je me dis chaque jour : Il faut que j'écrive à Norbert,mais il y a des dizaines d'années que cela dure ainsi.

L'été s'est à la fois trés bien et trés mal passé.Bien passé car agité par les passions! Mal passé car il pleuvait à Navarrenx pendant des semaines entiéres et j'ai fini par avoir des embêtements sérieux du côté des poumons.

Et toi? J'espére que tu vas te précipiter sur ta machine à écrire pour me dire comment tu vas.J'espére toujours que tu vas m'annoncer ton arrivée en France.Mais il faudrait me l'annoncer à l'avance pour que je prépare la reception et le séjour.

Il est aussi possible que j'aile te voir .J'aurai du venir à Ottawa en fin septembre.J'ai annulé pour diverses raisons dont la raison de s santé.Je pense que j'attendrai le printemps pour revenir dans ces pays où les frileux redoutent l'hiver.

Je dicte cette lettre à une trés aimable fille que j'adore et qui a trés envie d'abord de faire ta connaissance et ensuite de visiter les Amériques.Si j'arrive au printemps,au moment de la fonte des neiges comme cette année,j'espére que ce sera avec elle.

Voici quelques autres nouvelles.Armelle est en pension sur sa demande non loin de Paris.Nicole va avoir un bébé c'est à dire qu'elle s'insta lle dans sa nouvelle vie conjugale.Quant à Charlotte de plus en plus identique à elle même.Pour tout le reste ,consulte la presse.

Lefebvre to Guterman, October 18, 1977: "The Youthfulness of Heart" ["la Jeunesse du Coeur"]. Norbert Guterman Papers, Rare Book and Manuscript Library, Columbia University. Reproduced by permission of Columbia University and Moira Hyle.

(Figure continued on next page.)

J'allais oublier de te dire que catherine (la dactylo)et moi(le parleur) faisons un livre ensemble:Une suite de dialogues entrexxxx philosophiques et politiques entre une trés jeune femme et un monsieu qui n'a plus que la jeunesse du coeur....

C'est tout pour le moment...

Armelle aurait du écrire à Marguerite Sygminton pour la remercier du trés joli bijou. .Je ne sais si elle l'a fait.Il me semble qu'elle a préparé à Navarrenx une boite avec des cadeaux ,je ne sais si elle l'a envoyée.Cette fille ne sait plus écrire une lettre,elle ne sait que téléphoner.Elle n'est pas la seule.

Amitiés à tous, affectueusement à toi.

H. Lefebvre

real estate and financial assets, liquid loot yearning to become concrete in space.

"It's vital," says Lefebvre, "to underline the role of urbanism and more generally 'real estate' (speculation, construction) in neocapitalist society. 'Real estate,' as one calls it, plays a role of a secondary sector, of a circuit parallel to that of industrial production, which serves a market for 'goods' nondurable or less durable than 'buildings.' " "This secondary sector," he believes, "absorbs shocks" (p. 211; p. 159). In a depression, capital flows toward it, resulting in fabulous profits at first, profits that soon get fixed and tied down in the built environment, literally imprisoned in space (p. 211; p. 159). Capital fixates itself (*s'immobilise*) in real estate (*immobilier*), and soon the general economy begins to suffer. Yet the secondary circuit of capital expands all the same. Speculation assumes a life of its own, becomes at once enabling and destabilizing, a facilitator as well as a fetter for economic growth over the long term. "As the principal circuit, that of industrial production, backs off from expansion and flows into 'property,' " Lefebvre cautions, "capital invests in the secondary sector of real estate. Speculation henceforth becomes the principal source, the almost-exclusive arena of formation and realization of surplus value. Whereas the proportion of global surplus value amassed and realized in industry declines, the amount of surplus value created and realized in speculation and property construction increases. The secondary circuit thus supplants the primary circuit and by dent becomes essential" (p. 212; p. 160).

Lefebvre, as ever, never backs up this hypothesis with empirical data and insists often it's a "virtual object" he's constructing. But the speculative monomania within our own economy, kindled during the deregulated 1980s, and the emergence of the entrepreneurial city—where urban fates and fortunes are inextricably tied to the dynamics of stock markets—are all too evident. Banks, finance institutions, big property companies, and realtors spearhead the formation of a secondary circuit. Here capital circulates in pursuit of higher rental returns and elevated land prices. If ground rents and property prices are rising and offer better rates of return than other industrial sectors, and if finance is available at affordable interest rates, capital sloshes into assorted "portfolios" of property speculation. Cleaner and faster profits are in the offing. From capital's point of view as a class, this makes perfect bottom-line sense: the landscape gets flagged out as a pure exchange value, and activities on land conform to the "highest," if not necessarily the "best," land uses. Profitable locations get pillaged as the secondary circuit flows becomes torrential, just as other sectors and places are asphyxiated through disinvestment. Willy-nilly people are forced to follow hot money, flow from the countryside into the city, from factories into services, from stability into fragility. The urban fabric wavers between devaluation and revaluation, crisis and speculative binge, a ravaged built form and a renewed built form—and a fresh basis for capital accumulation. Once, it was a gritty warehouse or a rusty wharf; now, it's a glitzy loft or a prim promenade. Once, it was an empty field on the edge; now it's core neighborhood on the up.

This tendency was likewise spotted almost around the same time by Lefebvre's Anglo-Saxon soul mate, David Harvey. Near the end of *Social Justice and the City* (1973), in his "Conclusions and Reflections," Harvey rues that his seminal urban text was completed before he'd had the opportunity to study Lefebvre's *La révolution urbaine*.[2] "There are parallels between his concerns and mine," Harvey admitted, "and there are similarities in interpretation in content (which is reassuring) and some differences in interpretation and emphasis (which is challenging)." Lefebvre's emphasis, said Harvey, "is more general than my own. … Nevertheless, I feel more confident in appealing to both Lefebvre's work and the material collected in this volume, in attempting to fashion some general conclusions concerning the nature of urbanism."[3] It was around the idea of an "urban revolution," orbiting within a "secondary circuit of capital," where Harvey bonded with Lefebvre. "Lefebvre makes a simplistic but quite useful distinction between two circuits in the circulation of surplus value," says Harvey (p. 312). However, the contention that the secondary circuit *supplants* the principal circuit "requires," Harvey notes, "some consideration" (p. 312).

Since *Social Justice and the City,* Harvey has refined and deepened a lot of these ideas, because and in spite of Lefebvre. "The Urban Process under Capitalism: A Framework for Analysis," for example, is a brilliant reinterpretation of the "secondary circuit of capital" thesis and remains a classic Marxist disquisition.[4] Another early, thought-provoking effort titled "Class-Monopoly Rent, Finance and the Urban Revolution" equally acknowledges its debt to Lefebvre, while turning up the analytical heat several degrees. "Distinctions between 'land' and 'capital' and between 'rent' and 'profit,' " Harvey argues, "have become blurred under the impact of urbanization." "It may be," he continues, "that problems of 'stagflation' in advanced capitalist countries are connected to the land and property boom evident since the mid-1960s." Thus, urbanization has changed "from an expression of the needs of industrial producers to an expression of the power of finance capital over the totality of the production process."[5] (Harvey recently commented on the English translation of *The Urban Revolution* in a noteworthy review essay. "Rereading it anew," he admits, after first encountering the book in 1972, "turned out to be much more than a trip down memory lane. The text has lost none of its freshness, its beguiling and tantalizing formulations. The questions it opens up are still with us and deserve a thorough airing.")[6]

In 1973, Harvey thought Lefebvre pushed things too far, argued too prematurely in favor of urbanization. "The two circuits are fundamental to each other, but that based on industrial capitalism still dominates" (p. 313). Notwithstanding, "to say that the thesis is not true at this juncture in history," reckons Harvey, "is not to say that it is not in the process of becoming true or that it cannot become true in the future" (p. 313). Lefebvre insists that positing the urban over the industrial begets a new sort of "urban problem," which imposes itself globally and locally—and ideologically. "Urban reality," he says, "modifies the relations of production, without sufficing to transform them. It becomes a productive force, like science. Space and the politics of space 'express' social relations, but equally react back on them" (p. 25; p. 15).

For those "specialists" involved in urban problems, these circumstances become what Lefebvre coins a "blind field," something out-of-sight and out-of-mind. "Inasmuch as we look attentively at this new field, the urban, we see it with eyes, with concepts, shaped by the theory and practice of industrialization, with analytical thought fragmented and specialized in the course of an industrial epoch, thus *reductive* of the reality in formation" (p. 43; p. 29). From a Marxist perspective, a new dialectical reevaluation is called for, a revised theory of commodity production and surplus value extraction, a new spin on questions of class and economic growth. Indeed, at a time when dominant strands of Marxism—like Althusser's structuralism—were "formalizing" Marxism, hollowing out its content, Lefebvre was adamant that "urban reality" wasn't "superstructural," wasn't epiphenomenal to productive industrial forces, to the production of "tangible" commodities.

* * *

At the beginning of the 1970s—"the repugnant seventies," as Guy Debord likened them—Lefebvre penned numerous diatribes contra Althusser's structural Marxism and against structuralism more generally. Texts like *L'idéologie structuraliste* (1975) should be read alongside *The Urban Revolution,* for, in Lefebvre's eyes, the reign of structuralism chimed nicely with the state's *structuration* of urban reality. Structuralism's preoccupation with "system" and "systematization," he says (*L'idéologie structuraliste,* p. 70), "dehydrates the lived" and ends up as an ideological apologia for a bureaucracy it often sought to critique.[7] Here Althusser is the target of Lefebvre's frontal, yet Claude Levi-Strauss and Michel Foucault are also mauled. "In the developments of May '68," Lefebvre points out, "the student avant-garde rejected the dogmatic arrogance of structuralist tendencies, which, with the force of 'scientific' arguments, refuted the spontaneity of the insurgents. ... Afterwards, structuralist dogma regained its gravity, a cold allure baptized 'serious' and 'rigorous,' the allure of neo-scientism. It isn't only that this scientism (which purports to be pure under the epistemological break) neglects 'real' problems and processes; it also withdraws into a Fortress of Knowledge it never exits. During this same period, the bureaucratic state *structured* efficiently the whole world."[8]

In *The Urban Revolution,* Lefebvre promulgates a "formal" schema of his own, dialectically formulated, throwing light on the complex "levels and dimensions" of this new geographic, economic, and political reality. As urbanization annihilates time and space, connecting hitherto unconnected parts of the globe, tearing down barriers and borders in its market intercourse, using technology to speed up production and circulation, Lefebvre, like Marx in the *Grundrisse,* mobilizes abstraction to pinpoint the various moments of this "unity of process." The schema operates in *four* dimensions, a device that shifts temporally while it stretches out spatially, onto a global scale with height and breadth and everyday depth. He distinguishes a global level (G), where power is exercised and accommodates the most abstract relations, like capital markets and spatial management; an everyday *lived* level, the private (*privée*) (P) scale of *habiter;* and a *meso,* intermediate level, the urban scale, that incorporates and mediates between the global and the private and is hence "mixed" (M).

The global level is the realm of abstract power and the state, whose will is exercised through some kind of representation, usually of politicians and men of means who assert themselves strategically. "We know today," Lefebvre claims, "that in capitalist society two principle strategies are in use: *neoliberalism* (which maximizes the amount of initiatives allowed to private enterprise and, with respect to 'urbanism,' to developers and bankers); and *neomanagerialism,* with its emphasis (at least superficially) on planning, and, in the urban domain, on the intervention of specialists and technocrats and state capitalism. We also know there are compromises: neoliberalism leaves a certain amount of space for the 'public sector' and activities by government services, while neomanagerialism cautiously encroaches on the 'private sector' " (p. 107; p. 78).

It's at the meso, urban level (M), though, where all this comes together, where an abstract global reach attains everyday coherence. The "specifically urban ensemble," Lefebvre notes (p. 109; p. 80), "provides the characteristic unity of the social 'real.' " As such, the M level has a "dual purpose": on one hand, there's what happens *in* the city, within its internal relations and jurisdiction, within its built (and unbuilt) environment, within its private households (P); on the other hand, there's what happens *of* the city, its connectivity to surrounding areas, to other cities and spaces, and to its global hinterlands (G). "Lived" reality (P) functions within a *regime* of global capital accumulation (G) and a *mode* of state regulation mediated at the meso, urban domain (M).[9] This mixed urban scale becomes both the springboard for global mastery and the deadweight crushing the everyday. At the same time, a new hybrid Frankenstein is at the helm: the neoliberal bureaucrat and the managerialist entrepreneur, who embrace one another on the threshold of late capitalist urban change and global transformation.[10] Lefebvre says these managers and strategists, bankers and bureaucrats, politicians and pinstripes project themselves onto a global canvas as they colonize the lived. And they unite around a common urban *praxis:* "the generalized terrorism of the quantifiable" (p. 244; p. 185).

This motley band he pejoratively dubs "the urbanists," who "cut into grids and squares." "Technocrats," Lefebvre notes, "unaware of what's going on in their own mind and in their working concepts, profoundly misjudging in their blind field what's going on (and what isn't), end up meticulously organizing a repressive space" (p. 208; p. 157). Urbanism thus finds itself caught between the rock and the hard place, "between those who decide on behalf of 'private' interests and those who decide on behalf of higher institutions and power." The urban wilts under a historic compromise between neoliberalism and neomanagerialism, "which opens the playing field for the activity of 'free enterprise.' " The urbanist duly slips into the cracks, making a career in the shady recesses between "developers and power structures," a monkey to each organ grinder. A true left critique, accordingly, must attack the promoters of the urban "as object," as an entity of economic expansion in which investment and growth are ends in themselves. The agents of this mind-set, meanwhile, the top-down, self-perpetuating *cybernanthropes,* must everywhere and always be refuted.

For Lefebvre, the cybernanthrope was the antihumanist incarnate, a reviled man cum machine, the air-conditioned official obsessed with information systems, with scientific rationality, with classification and control. In a profoundly witty and scathing text, *Vers le cybernanthrope* [Towards the cybernanthrope] (1971), Lefebvre claims cybernetic culture has cut—not unlike Robert Moses slicing into New York—a swath for the urban revolution and proliferated through urbanism as ideology. *Voici* everything Lefebvre hates. Their type, their policies, their urban programs, the very presence of technocrats on planet earth offended him; they were antithetical to all he stood for, all he desired. Their type plots in think tanks and research units, in universities and in chambers of commerce, discourse with PowerPoint and flip charts in boardrooms near you, formulate spreadsheets and efficiency tables, populate government and peddle greed. They thrive off audits and evaluation exercises, love boxes, and ticking off numbers. Their political remit and strategic program reaches supragovernmental status these days in the citadels of the World Bank, International Monetary Fund, and World Trade Organization, where "good business" dictates and "structural adjustment" initiatives become carrots and sticks in urban and global "best practices."

Yet in *Vers le cybernanthrope*, Lefebvre is more ironic than irate. The cybernanthrope enforces himself as a "practical systematizer," he says, determining those boundaries socially permissible, stipulating order and norms, conceiving "efficiency models," and organizing equilibrium, *feedbacks,* and homeostasis. "The cybernanthrope deplores human weakness," Lefebvre thinks. "He disqualifies humanism in thinking and action. He purges the illusions of subjectivity: creativity, happiness, passion are as hollow as they are forgettable. The cybernanthrope aspires to function, to be the only function. … He's a man who receives promotion and lives in close proximity with the machine," be it laptop or desktop.[11]

> He adheres to a cult of equilibrium in general and to his own in particular, protecting it intelligently. He aims to maintain stability, to defend it. The principles of economics and a minimum of action are his ethical principles. The cybernanthrope ignores desires. Or if he recognizes desire, it's only to study it. There are only needs, clear and direct needs. He despises drunkenness. As an Apollonian, the Dionysian is a stranger to him. The cybernanthrope is well nourished and smartly dressed. He mistrusts unknown flavors, tastes too rich or too surprising. Odors—they're something incongruous, incontrollable, archaic. What pleases him most is to have everything pasteurized, everything hygienic and deodorized. He treats severely

> the dramatic, the historic, the dialectic, the imaginary, the possible–impossible. Anything that doesn't reveal itself in rationality, or in his programmatic discourse, is rejected as folklore.[12]

Little wonder these guys followed Le Corbusier's wisdom and "killed the street." The cybernanthropic urbanist maintains that streets are "traffic machines" where the "object-king" or "object-pilot" circulates, where vehicles imbued with surplus value shift commodities and labor power. "The invasion of the automobile," Lefebvre says in *The Urban Revolution,* "and the pressure of this industry and its lobbyists … have destroyed all social and urban life" (p. 29; p. 18). "When you eliminate the street, there are consequences: the extinction of all life, the reduction of the city to a dormitory, to an aberrant functionalization of existence." But the street "contains qualities ignored by Le Corbusier" (p. 30; p. 18). In the street, there's an informative, symbolic, and ludic function. In the street, you play and you learn stuff. "Sure, the street is full of uncertainty. All the elements of urban life, elsewhere congealed in a fixed and redundant order, liberate themselves and gush onto the street and flow towards the center, where they meet and interact, freed from fixed moorings" (p. 30; p. 19). In the street, "disorder lives. It informs. It surprises." However, in the street, he says, concurring with Jane Jacobs, whom he cites approvingly, this disorder constructs a superior order (p. 30; p. 19).

That other sort of revolution unfurls in the street, Lefebvre reminds his Marxist readers, in case any are listening. "Doesn't this likewise illustrate that disorder engenders another order? Isn't the urban street a place of speech, a site of words much more than of things? Isn't it a privileged domain where speech is scripted? Where words can become 'wild,' daubed on walls that elude rules and institutions?" (p. 30; p. 19). In the street, everyday tongues converse in *argot* rather than discourse in jargon, the remit of cybernanthropes and specialists, who conceive in offices rather than occupy streets.[13] In the street, you find rough talk, raw energy, profanities that disrupt and unnerve the cybernanthrope. The impulse for revolt will come from the street, Lefebvre knows. There, a new "style" will bloom, vanquishing the cybernanthrope, overcoming his *faux* urbanism.[14] This style will affirm the *anthrope,* a humanist nemesis, armed to the teeth with weapons of irony and humor, art and literature. "The war of anthropes contra cybernanthropes," he says, "will be a guerrilla war. Anthropes will have to elaborate a strategy founded upon the destruction of the cybernanthrope's order and equilibrium."[15] "For vanquishing, or even for engaging in battle, anthropes should valorize imperfections: disequilibrium, troubles, oversights, gaps, excess and defects of consciousness, derailments, desires, passion and irony. The anthrope should always fight against a plan of logic, of technical perfection, of formal rigor, of functions and structures. Around rocks of equilibrium will be waves and air, elements that will erode and reclaim."[16]

* * *

Revolutionary refrains emanating from below, from a street praxis, are admittedly hushed in *The Urban Revolution*. Lefebvre has given us a quieter, more reflective analytical text, more cautious in its militant musings. But the idea of "vanquishing by style" offers clues to his revolution hopes, even if they're now dimmer. Here, for guidance, we must turn back the clocks briefly, to a pre-1968 work, *La Proclamation de la Commune,* written in 1965. It's hard to decide whether Lefebvre's subject matter here was 1871 or 1968—whether he was excavating the past or foreseeing the future; whether this was a historic day in March 1871, shattering the Second Empire, reclaiming Paris's center for the people, toppling the imperial mantle of Napoleon III and sidekick Baron Haussmann or an imminent student–worker eruption that would almost smash the Fifth Republic of de Gaulle. Either way, it was the *style* of the Commune that whetted Lefebvre's political palate. The Commune's style, he says, "was, first of all, an immense, grandiose festival, a festival that citizens of Paris, essence and symbol of the French people and of people in general, offered to themselves and to the world. Festival at springtime, festival of the disinherited, revolutionary festival and festival of revolution, free festival, the grandest of modern times, unfurls itself for the first time in all its dramatic magnificent joy."[17]

For seventy-three days, loosely affiliated citizen organizations, neighborhood committees, and artist associations converted Paris into a liberated zone of anarcho-socialism. It was, Lefebvre notes, "grandeur and folly, heroic courage and irresponsibility, delirium and reason, exaltation and illusion" all rolled into one.[18] Insurgents somehow corroborated Marx's notion of revolutionary praxis at the same time as they refuted it, for this was as much a geographical as a historical event, no worker uprising incubated in the factories; rather, it was "the grand and supreme attempt of a city raising itself to the measure of a human reality."[19] An urban revolution had made its glorious debut, reenergizing public spaces and transforming everyday life, touting victory while it wobbled in defeat. It was condemned to death at birth, despite the gaiety of its baptism. "The success of revolutionary movement," Lefebvre says, "masked its failings; conversely, its failures are also victories, openings on to the future, a standard to be seized, a truth to be maintained. What was impossible for the Communards stays until this day impossible, and, by consequence, behooves us to realize its possibility."[20] "We are thus compelled," he reasons, "to rehabilitate the dream, otherwise utopian, and put to the forefront its *poetry,* the renewed idea of a *creative praxis*. There resides the experience of the Commune and its style."[21]

This rhetorical flourish lingers in *The Urban Revolution*. But there it takes on a new twist, has an even broader message and implication. The urbanism of Haussmann tore out the heart of old medieval Paris and reinvented the concept of a center, of a downtown of bright lights and conspicuous consumption. Erstwhile pesky proletarians would take hold of shovels, man the building sites, and have no time to make trouble. They'd also find themselves dispatched to a rapidly expanding *banlieue,* to the new suburbs mushrooming in the distance. In one sense, Paris gained as an independent work of art, as an aesthetic experience admired to this day by every tourist and visitor. Yet in another sense it lost something as a living democratic organism, as a source of generalized liberty. Hence Haussmann not only patented what we'd now call the gentrified city, with its commodification of space, but also pioneered a new class practice, bankrolled by the state: the deportation of the working class to the periphery, a divide-and-rule policy through urbanization itself, gutting the city according to a rational economic and political plan. The logic of the city would never quite be the same again.

While we can quibble with Lefebvre over that class practice stateside, where, aside from a few exceptions (Manhattan, Boston, and San Francisco, etc.), the rich have decanted themselves to the periphery, bestowing on the poor an abandoned core, Lefebvre's point is more global in scope. The urban revolution is now a "planetary phenomenon," he says; urbanization has conquered the whole world, left nowhere unscathed, nowhere "pure" anymore. To that degree, Haussmannization is now a *global class practice,* an urban strategy that peripheralizes millions and millions of people everywhere. As cities explode into megacities and as urban centers—even in the poorest countries—get glitzy and internationalized, "Bonapartism" (as Lefebvre coins it) projects its urban tradition onto the twenty-first-century global space. The peripheralization of the world's least well off is apace. By 2020, we can repeat, two billion will inhabit *favelas* and *bidonvilles* scattered around the edge of the world's biggest cities. By 2015, nineteen of the twenty-three boomtowns predicted to have populations in excess of ten million will be in "developing" countries. The vast global suburb in the making will thus be homemade, teetering in the breeze, waiting for the perfect storm.

"Can such a strategy assume that the countryside will invade the city," Lefebvre asks, "that peasant guerillas will lead the assault on urban centers?" (p. 152; p. 113). Régis Debray's *Revolution in the Revolution?* had voiced this thesis a few years earlier, in 1967, and Lefebvre seems to want to respond. The city put the brake on revolutionary momentum, Debray said, was a hypertrophic "head," full of abstract ideas, deaf to the plight of peasant guerillas; the rural hinterlands and mountain jungles were the "armed fist" of the liberation front. The city corrupts radicalism, made comrades soft and lulled them into the trappings of bourgeois life. "The mountain proletarianizes bourgeois and peasant elements while the city bourgeoisifies proletarians."[22]

"Today," Lefebvre counters, "such a vision of class struggle on a global scale appears outdated. The revolutionary capacity of the peasantry is not on the rise" (p. 152; p. 113). In fact, it is being "reabsorbed" within an overall colonization of space, where both peasants and proletarians occupy not rural hinterlands but urban hinterlands, each marginalized at the urban periphery, out on the *world-city banlieue*. A global ruling class, meanwhile, shapes out its core, at the center, Haussmannizing nodes of wealth and information, knowledge and power, creating a feudal dependency within urban life. "In this case," concludes Lefebvre, "the frontier line doesn't pass between the city and the country, but is within the interior of the phenomenon of the urban, between a dominated periphery and a dominating center" (p. 152; p. 113).[23] In a word, the urban revolution is as much ruralization of the city as urbanization of the countryside.

Will those two billion dispossessed ever want to stake a claim to the core, assert their right to centrality, *de*marginalize themselves with grand style, in a giant street festival? Will globalization of communication and publicity open everything up to "the eyes of the global poor"—adapting Baudelaire's poem—inspiring indignation and organization as well as awe ("big saucers eyes"), prompting the "world literature" Marx dreamed of in *The Communist Manifesto*? Tens of thousands of poor landless Latinos have already helped reinvent the urban labor movement in California; militancy in South African townships brought down Apartheid; millions took to the streets in Jakarta, Seoul, Bangkok, São Paulo, and Buena Aires, when East Asian and Latin American economies went into meltdown during 1997; revolts against the International Monetary Fund shock therapy programs have regularly left many developing world capitals smoldering as the most vulnerable connect the global with the local on the street. Examples abound. The fault line between the internationalization of the economy and a marginalization of everyday life scars urban space. The urban scale is the key mediator on the global scene, at once the stake and terrain of social struggle, both launch pad and linchpin in history. The urban revolution from below, as a historic bloc—or seismic tremor—still remains the "virtual object" Lefebvre described in 1970, a future scenario yet to be established. But if it ever becomes a "real" object, a directly lived reality, insurgency will look a lot different from 1968 and 1871, and from 1917: the storming of the Winter Palace will now come in the monsoon season.

* * *

The Urban Revolution intimates the shape of things to come, in terms of both Lefebvre's political desires and his scholarly œuvre. As ever, one project prompted another, one thesis led him to an antithesis and consequently to a higher thesis, something to be confirmed, tested out, negated again. His life and work ceaselessly moved through this dialectical process of affirmation and negation, thesis and antithesis and synthesis. In one footnote, buried deep within *The Urban Revolution* (p. 179; pp. 194–95), he says he'll be returning to some of the book's contents in another monograph, bearing the title *Theories of Urban Space*. Lefebvrian aficionados will know this text never materialized under that rubric. They'll know instead that here lie the seeds of what would eventually become *The Production of Space,* regarded by many critics as his most accomplished work.

More and more Lefebvre believes that despite their "blind fields," technocrats and cybernanthropes did see with collective clarity when it came to one aspect of neocapitalist reality: they knew that executing their will meant obeying a "social command." This writ wasn't accented on such and such a thing, on such and such an object, as on a "global object, a supreme product, the ultimate object of exchange: space" (p. 204; p. 154). Thus this power to control future economic and political destinies is predicated on a command not of objects in space but of space itself. "Today," Lefebvre says, "the social (global) character of productive labor, embodied in the productive forces, is apparent in the social production of space" (p. 205; p. 155).

> Today, space as a whole enters into production, as a product, through buying and selling and the part exchange of space. Not too long ago, a localized, identifiable space, the soil, still belonged to a sacred entity: the earth. It belonged to that cursed, and therefore sacred, character (not the means of production but the Home), a carryover from feudal times. Today, this ideology and corresponding practice is collapsing. Something new is happening. (p. 205; p. 155)

6

SPACE

I hear the ruin of all space, shattered glass and toppling masonry.

—James Joyce, *Ulysses*

The Production of Space was Henri Lefebvre's fifty-seventh book, the crowning glory of research on cities and spatial questions, spanning the 1968–74 period, when, aside from lecturing and witnessing students go into revolt mode, he scribed nine books and a dozen articles and helped found the journal *Espace et société*.[1] To write *The Production of Space,* emeritus-to-be Henri was given a special stipend from Paris-Nanterre, his employer, and the densely argued, 485-page tome was worth every centime, reigning as it does as one of his greatest and most enduring works. The book, it's equally been noted, was personally important to Lefebvre, because it punctuated the end of his truncated yet illustrious academic career.[2] If this spatial moment came in the twilight of Lefebvre's career, when it came it literally erupted. Just as the mature Karl Marx never chose political-economy as his vocation but rather political-economy chose him, space now seemed to choose Lefebvre as its critical conscience; it was the state of the world, as opposed to the state of his mind, that prompted his intellectual engagement, spurred his rejigging of the Marxist historical object, of a general theory of production that hitherto unfolded on the head of a pin.

* * *

Remi Hess has pointed out a curious Lefebvrian factoid. Despite being widely translated into scores of languages, there's a geography and temporality to the uptake of Lefebvre's books. His texts, in short, haven't all been translated in the same countries at the same time.[3] The Japanese have translated a lot, Anglo-Americans have translated a handful, notably since 1991; German reeditions, and those released in Latin American countries, have their own politically conditioned logic; ditto South Korea, who today is a big Lefebvrian importer, where his texts sell like radical hotcakes. Moreover, works deemed important by aficionados, like *Nationalisme contre les nations* (1937) and *La Somme et le Reste* (rereleased in France in 1989), haven't sold well. Many Lefebvre texts are simply out of print or perhaps out of fashion, even if they're never out of sync. (Hess pointed out that since the fall of the Berlin Wall, Lefebvre's Marxist œuvre has dwindled and become antiquarian stuff.) By the early 1990s, there wasn't a single Lefebvre book in print in France; his renown there had apparently receded from the public realm. Furthermore, sales of *La production de l'espace,* whose fourth edition appeared in 2000, stretch somewhere between three and four thousand copies, whereas *The Production of Space* now tops almost twenty thousand copies.[4]

Given this minority homeland status, why, we might wonder, has Lefebvre become an almost cult figure in Anglo-American critical–theoretical circles? Did his work on space initially lead to bewilderment in France? Maybe this spatial moment sounded the death knell of Lefebvre's intellectual acclaim? *The Production of Space* was misunderstood and overlooked when it hit French bookshelves in 1974. The timing couldn't have been worse: by then Althusser's reputation was formidable and his structural Marxism was de rigueur; he was the flagship of French theory's arrival across the Channel and across an ocean. And if you didn't agree with Althusser and you were still a Marxist, you'd turn to Roger Garaudy's humanism, not Lefebvre's. There was seemingly little intellectual scope for Hegelian Marxism.[5] And a book about space? That's what most socialist radicals seemed to need like a hole in the head! When things did assume an urban turn, in the early phases of *Espace et société,* Althusser still curiously snuck in ahead of Lefebvre. It was the former's Marxism, after all, that underwrote Manuel Castells's highly influential sociological research on urbanization: Castells's *La quéstion urbaine*—replete with attacks on former mentor Lefebvre—made it to press two years before *La production de l'espace* and undercut his senior's humanist predilections and analytical pretensions.

In fact, Castells asked whether the "urban" was a legitimate object of inquiry at all. The "urban question" for him was above all a question of how an urbanizing *capitalist mode of production* functioned. In Castells's spatial universe, the city was indeed a *container* of social and class relationships. But it was these social relations that had primacy over any explicit "urban" or "spatial" category. Lefebvre, for Castells, had strayed too far, had *reified* space; Castells caught a whiff of spatial fetishism, attributing to the spatial causal determinacy over the societal. From trying to develop a "Marxist analysis of the urban phenomenon," Lefebvre, Castells said, "comes closer and closer, through a rather curious intellectual evolution, to an *urbanistic theorization of the Marxist problematic*."[6] No compliment intended: this was a stinging rebuttal, probably helping ensure the relative neglect of Lefebvre's work during the 1970s.[7]

While Lefebvre's rejoinder maintained that Castells didn't understand space—"He sets aside space," Lefebvre scoffed. "His is still a simplistic Marxist schema"[8]—it was David Harvey who brought Lefebvre to the attention of Anglophone audiences. In *Social Justice and City,* we know, Lefebvre only cameoed. Yet his idea that a distinctively "urban revolution" was supplanting an "industrial revolution" and that this urban revolution was somehow a spatial revolution as well had a deep and lasting resonance in critical urban studies and geography—longer lasting, it seems, than Castells's own urban research, which was reaching its sell-by date as early as the mid-1980s. Steadily, from the mid-1970s onward, Lefebvre's urban and spatial ideas seeped into Anglophone urban and geographical scholarship, spawning, by the early 1980s, a Lefebvrian cottage industry of sociospatial Marxism. In this context, rather than Lefebvre influencing English-speaking geography and urbanism, it's perhaps been the other way around: maybe it has been Anglo-American spatial theorists who've resuscitated Lefebvre's flagging spatial career, prompted his more recent (posthumous) claim to fame. Michel Trebitsch, in his essay on Volume 3 of *Critique of Everyday Life,* forthcoming as a preface to Verso's English translation, even reckons this Anglophone Lefebvrian turn has reacted back into France, giving a "new look" to his œuvre there, "re-acclimating" it within "classic French theory."

One wonders how widespread Lefebvre's work would have been without the first-wave mediation of David Harvey (instrumental in pushing for an English translation of *La production de l'espace*), Ed Soja, Fredric Jameson, Mark Gottdiener, Derek Gregory, Kristin Ross, Elenore Kofman, and Elizabeth Lebas, as well as second-wave interpreters like Rob Shields, Erik Swyngedouw,

Stuart Elden, Stefan Kipfer, and Neil Brenner. One wonders, too, whether we'd have ever seen *The Production of Space* appear in English. God knows, seventeen years is a stretch anyway, a far cry from Althusser's *For Marx* (published in 1965 and making it into English a couple of years later). Debuting in 1991 and capably translated by one-time Brit Situ Donald Nicholson-Smith, *The Production of Space* has been the biggest catalyst in Lefebvre's rise to Anglophone stardom. Its appearance was *the* event within critical human geography during the 1990s, sparking a thorough reevaluation of social and spatial theory, just when apologists for a globalizing neoliberalism proclaimed "the end of geography." After a very long wait, English audiences not only have been given access to a classic text of Marxist geography, they've equally been living through the very productive process this book underscored.

* * *

The explorations in *The Production of Space* (POS in citations) are explorations of an extraordinary protean, seventy-three-year-old French Marxist. Of course, there's much more going on than plain old-fashioned Marxism: Hegel crops up often; Nietzsche's spirit is palpable; and Lefebvre's grasp of romantic poetry, modern art, and architecture is demonstrable. Meanwhile, he breezes through the history of Western philosophy as if it's kids' stuff, as if everybody understands his unreferenced allusions, his playful punning and pointed pillorying. Prominent here are the diverse "moments" within Lefebvre's own œuvre: his philosophical moment, his literary moment, his historical and political moments, plus a moment we can describe as a moment of *confrontation*. The book begins with a "Plan of the Present Work,"[9] an opening gambit of masterful coherence, whose argument proceeds with considerable analytical consistency and lucidity. Immediately, we get a compressed account of the concept of space, listen to how it has been denigrated in Western thought, within the Cartesian tradition, by Kant, by Bergson, and by structural linguistics, and hear how Lefebvre himself aims to tread through this mottled landscape. On the face of it, this all sounds like a tame philosophical dilemma, hardly one to change the world. But as we follow Lefebvre onward through *The Production of Space,* we soon see its radical import.

After a while, his pursuit for a "unitary theory of space" unfolds—critically and flamboyantly. The project he coins *spatiology* (POS, p. 404) and involves, among other things, a rapprochement between *physical* space (nature), *mental* space (formal abstractions about space), and *social* space (the space of human interaction). These different "fields" of space, Lefebvre says, have suffered at the hands of philosophers, scientists, and social scientists, not least because they've been apprehended as separate domains. *The Production of Space* seeks to "detonate" everything, to readdress the schisms and scions; Lefebvre considers fragmentation and conceptual dislocation as serving distinctly ideological ends. Separation ensures consent and perpetuates misunderstanding; or worse, it props up the status quo. By bringing these different "modalities" of space together, within a single theory, Lefebvre wants to *expose* and *decode* space, to update and expand Marx's notion of *production,* to leave the noisy sphere where everything takes place on the surface, in full view of everyone, and enter into the hidden abode, on whose threshold hangs the following notice: "No admittance except on business!"

The emphasis on production is, of course, very Marxist. To be radical, for Marx, meant "grasping things by the root."[10] And his obsession with production was designed to do just that: to get to the root of capitalist society, to get beyond the fetishisms of observable appearance, to trace out its inner dynamics and internal contradictions, holistically and historically. Lefebvre likewise demystifies capitalist social space by tracing out its inner dynamics and generative moments—in all their various physical and mental guises, in all their material and political obfuscations. Here, *generative* means "active" and "creative," and *creation,* he says, "is, in fact, a *process*" (POS, p. 34). Thus, getting at this generative aspect of space necessitates exploring how space gets *actively produced.* Again, like Marx in his theoretical quest for explanation, Lefebvre makes political purchase of process thinking, of conceiving reality in *fluid movement,* in its *momentary existence* and *transient nature.*

Now, space becomes reinterpreted not as a dead, inert thing or object but as organic and alive: space has a pulse, and it palpitates, flows, and collides with other spaces. Lefebvre's favorite metaphors hail from hydrodynamics: spaces are described in terms of "great movements, vast rhythms, immense waves—these all collide and 'interfere' with one another; lesser movements, on the other hand, interpenetrate" (POS, p. 87). "All these spaces," he adds, "are traversed by myriad currents. The hyper-complexity of space should now be apparent, embracing as it does individual entities and particularities, relatively fixed points, movements, and flows and waves—some interpenetrating, others in conflict" (POS, p. 88). And these interpenetrations—many with different temporalities—get superimposed on one another in a *present* space; different layers of time are inscribed in the built landscape, literally piled on top of each other, intersecting and buried, palpable and distorted within three-dimensional "objective" forms that speak a flattened, one-dimensional truth. Thus, "it's never easy," Lefebvre warns, "to get back from the object to the activity that produced and/or created it" (POS, p. 113). Indeed, once "the construction is completed, the scaffold is taken down; likewise, the fate of an author's rough draft is to be torn up and tossed away" (POS, p. 113). Revisiting an abandoned construction site, delving into the wastebasket of history, retrieving a crumbled draft are henceforth tantamount to "reconstituting the process of its genesis and the development of its meaning."

Here we have a vivid demonstration of Lefebvre's "regressive–progressive method," as well as a spatialized rendering of Marx's famous analysis on "the fetishism of commodities." From the present, from an actual predicament, Lefebvre's approach shifts backward, excavates the past, conceptually retraces it, burrows into grassed-over earth, then propels itself forward again, pushing onward into the frontiers of the virtual, into the yet-to-be. The production of space, he says, "having attained the conceptual and linguistic levels, acts retroactively upon the past, disclosing aspects and moments of it hitherto misapprehended. The past appears in a different light, and hence the process whereby that past becomes the present also takes on another aspect" (POS, p. 65). Ditto for Marx, who moved backward from a "thing-like" entity, the commodity-form, whose development was most pronounced in mid-nineteenth-century England, to reconstruct the totality of capitalism's past and possible future.

The commodity, Marx said, possesses a "mystical" and "mist-enveloped" quality he labels "fetishism." At the marketplace, at the level of exchange—in a department store, a car salesroom, at The Gap—it's impossible to apprehend the activities and exploitations occurring in a productive labor process. What are fundamentally intersubjective relations become, Marx says, perceived by people as objective, as "a definite social relation between men that assumes, in their eyes, the fantastic form of a relation between things."[11] Lefebvre's epistemological shift, from conceiving "things in space" to that of the actual "production of space" itself, is the same quantum leap Marx made in his colossal, all-incorporating analysis of the capitalist mode of production:

> Instead of uncovering the social relationships (including class relationships) that are latent in spaces, instead of concentrating our attention on the production of space and the social relationships inherent to it—relationships which introduce specific contradictions into production, so echoing the contradiction

> between private ownership of the means of production and the social character of the productive forces—we fall into the trap of treating space "in itself," as space as such. We come to think in terms of spatiality, and so fetishize space in a way reminiscent of the old fetishism of commodities, where the trap lay in exchange, and the error was to consider "things" in isolation, as "things in themselves." (POS, p. 90)

Now, space is no more a passive surface, a *tabula rasa* that enables things to "take place" and action to ground itself somewhere; space, like other commodities, is *itself actively produced*: it isn't merely the staging of the theater of life as a paid-up member of the cast. Indeed, it's an "active moment" in social reality, something produced before it is reproduced, created according to definite laws, conditioned by "a definite stage of social development" (as Marx said in his *Grundrisse* introduction). Each mode of production has its own particular space, and "the shift from one mode to another must entail the production of a new space" (POS, p. 46); industrial capitalism dismantled feudal space, late capitalism has produced—goes on producing—its historically specific urban and industrial forms, continuing to colonize and commodify space, to buy and sell it, create and tear it down, use and abuse it, speculate on and war over it. The history of bourgeois geography is a historical geography of expropriation, both of property and of peoples, resounding with shattering glass and toppling masonry; it's written in the annals of civilization in letters of blood and fire.

Capitalism seemed to exhaust a lot of productive capacity, a lot of profitable capability in the postwar era: where to turn, what to do, who to exploit, and what to rip off? The system found new inspiration in the conquest of space. Not stratospheric space but human space, our everyday universe, with new *grands projets* on terra firma, transforming city cores and suburban peripheries, frontiers between countries, communications infrastructure; implanting new transcontinental networks of exchange within an emergent world market. To that degree, says Lefebvre, capitalism has bought time for itself out of the space it captures, out of the geographical niches it has created, the physical and social environment it absorbs. It has not resolved its inner contradictions as much as internalized them, displaced them elsewhere, broadened and deepened them. Contradictions of capitalism henceforth manifest themselves as contradictions *of* space. To know how and what space internalizes is to learn how to produce something better, is to learn how to produce another city, another space, a space for and of socialism. To change life is to change space; to change space is to change life. Neither can be avoided. This is Lefebvre's radiant dream, the virtual object of his concrete utopia. It's a dream that undergirds *The Production of Space*.

* * *

Critical knowledge has to capture in thought the actual process of production of space. This is the upshot of Lefebvre's message. Theory must render intelligible qualities of space that are both perceptible and imperceptible to the senses. It's a task that necessitates both empirical and theoretical research, and it's likely to be difficult. It will doubtless involve careful excavation and reconstruction; warrant induction and deduction; journey between the concrete and the abstract, between the local and the global, between self and society, between what's possible and what's impossible. Theory must trace out the actual dynamics and complex interplay of space itself—of buildings and monuments, of neighborhoods and cities, of nations and continents—exposing and decoding those multifarious invisible processes, as well as those visible practices of brute force and structural injustice. But how can this be done?

Lefebvre works through these dilemmas by constructing a complex heuristic: he calls it a "spatial triad," and it forms the weight-bearing epistemological pillar of *The Production of Space.* Unfortunately—or perhaps fortunately—he sketches this out only in preliminary fashion, leaving us to add our own flesh, our own content, to rewrite it as part of our own chapter or research agenda. What's more, while Lefebvre notes that the triad is something we'll encounter "over and over again" in *The Production of Space,* its appearance beyond the opening chapter is more implicit than explicit, assumed rather than affirmed. Why? Because it's no mechanical framework or typology he's bequeathed but a dialectical simplification, fluid and alive, with three specific moments that blur into each other: representations of space, spaces of representation, and spatial practices.

Representations of space refer to conceptualized space, to the space constructed by assorted professionals and technocrats. The list might include planners and engineers, developers and architects, urbanists and geographers, and others of a scientific or bureaucratic bent. This space comprises the various arcane signs and jargon, objectified plans and paradigms used by these agents and institutions. Representation implies the world of abstraction, what's in the head rather than in the body. Lefebvre says this is always a *conceived* space; usually ideology, power, and knowledge lurk within its representation. It's the dominant space of any society, "intimately tied to relations of production and to the 'order' those relations impose, and hence to knowledge, to signs, to codes, to 'frontal' relations" (POS, p. 33). Because this is the space of capital, state, and bourgeoisie, representations of space play a "substantial role and specific influence in the production of space" (POS, p. 42), finding "objective expression" in monuments and towers, in factories and office blocks, in the "bureaucratic and political authoritarianism immanent to a repressive space" (POS, p. 49).

Spaces of representation are directly lived spaces, the space of everyday experience. They are the nonspecialist world of argot rather than jargon, symbols, and images of "inhabitants" and "users" and "overlay physical space, making symbolic use of its objects" (p. 39). Spaces of representation are the café on the corner, the block facing the park, the third street on the right after the Cedar Tavern, near the post office. Spaces of representation may equally be linked to underground and clandestine sides of life and don't obey rules of consistency or cohesiveness, and they don't involve too much head: they're felt more than thought. A space of representation is *alive:* "it speaks. It has an affective kernel or center: Ego, bed, bedroom, dwelling, house; or, square, church, graveyard. It embraces the loci of passion, of action and of lived situations, and thus immediately implies time. Consequently, it may be qualified in various ways: it may be directional, situational or relational, because it is essentially qualitative, fluid and dynamic" (POS, p. 42).[12] Lived space is elusive, so elusive that thought and conception want to master it, need to appropriate and dominate it.

Spatial practices are practices that "secrete" society's space; they propound and propose it, in a dialectical interaction. Spatial practices can be revealed by "deciphering" space and have close affinities with *perceived* space, to people's perceptions of the world, of their world, particularly its everyday ordinariness. Thus spatial practices structure lived reality, include routes and networks, patterns and interactions that connect places and people, images with reality, work with leisure. Perceptual "imageability" of places—monuments, distinctive landmarks, paths, natural or artificial boundaries (like rivers or highways)—aid or deter a person's sense of location and the manner in which a person acts. Spatial practices, says Lefebvre, embrace production and reproduction, conception and execution, the conceived as well as the lived; they somehow ensure societal cohesion, continuity, and what Lefebvre calls "spatial competence" (POS, p. 33).[13] Yet cohesiveness doesn't necessarily imply coherence, and Lefebvre is vague about how spatial practices *mediate* between the conceived and the lived, about how spatial practices keep representations of space and spaces of representation together, yet apart. One thing he's sure of, though, is that there are "three elements" here not two. It's not a simple binary between lived and conceived but a "triple determination": each instance internalizes and takes on meaning through other instances.

Relations between conceived–perceived–lived spaces aren't ever stable, nor should they be grasped artificially or linearly. But Lefebvre has been around enough to know that lived experience invariably gets crushed and vanquished by the conceived, by a conceived *abstract space,* by an objectified abstraction. In this sense, abstract space is the product—the *materialization*—of what is conceived, a space of representation generalized. This idea of "abstract" again has Marxian overtones: *abstract space* bears an uncanny resemblance to Marx's notion of *abstract labor,* even though Lefebvre ventures much further than Marx, for whom "abstract" operated as an explicitly temporal category. Marx, remember, held that qualitatively different (concrete) labor activities got reduced to one quantitative (abstract) measure: money. Making a shirt is the concrete labor of a tailor whose use value is sanctioned by the market price for shirts; that is, by its exchange value. At such a point, what was concrete, useful, and particular becomes abstract, money driven, and universal. Money becomes the common denominator of all concrete things, of every labor activity that creates commodities; Marx coined this kind of labor *abstract labor,* labor in general, value-producing toil that's intimately tied to the "law of value," to socially necessary labor *time.*

In no way does "abstract" imply a mental abstraction: abstract labor has very real social existence, just as exchange value does, just as interest rates and share prices do. Similarly, abstract space has real ontological status and gains objective expression in specific buildings, places, activities, and modes of market intercourse over and through space. Yet its underlying dynamic is conditioned by a logic that shows no *real* concern for qualitative difference. Its ultimate arbiter is value itself, whose universal measure (money) infuses abstract space. Here exigencies of banks and business centers, productive agglomerations and information highways, law and order all reign supreme—or try to. And while the bourgeoisie holds sway in its production and organization, abstract space tends to sweep everybody along, molding people and places in its image, incorporating peripheries as it peripheralizes centers, being at once deft and brutal, forging unity out of fragmentation. Lefebvre asks us to open our eyes, to visualize the world dialectically, to see how homogeneous abstract space manifests itself in a dislocated and dismembered landscape of capitalism, a global space pivoting around "uneven development" and pell-mell differentiation. "The space that homogenizes," he declares, "thus has nothing homogeneous about it" (POS, p. 308).

* * *

There's nothing obvious or transparent about abstract space; it cannot be reduced to a single strategy. Although its nature *is* a conspiracy of sorts, it isn't *just* a conspiracy. Within abstract space are subtle ideological and political machinations, which maintain a perpetual dialogue between its space and users, prompting compliance and "nonaggression" pacts. The quasi-legal authority of abstract space imposes "reciprocity" and "commonality" of use, just as "in the street," Lefebvre jokes, analogously, "each individual is supposed not to attack those he meets; anyone who transgresses this law is deemed guilty of a criminal act" (POS, p. 56). You instinctively know your place, instinctively know where things belong; this intricate microfunction pervades abstract space's macrodetermination. Abstract space impregnates people, socializes everybody as spatial bodies and class subjects; its inbuilt consensus principle allows it to function within lived space and to flourish as all there is to be perceived. Just as abstract labor denies true concrete labor, renders labor without a market superfluous, abstract space ultimately denies concrete qualitative space: it denies the generalization of what Lefebvre calls *differential space,* the space of what socialism *ought* to be, a space that doesn't look superficially different but that is different, different to its very core. It's different because it celebrates bodily and experiential particularity, as well as the nonnegotiable "right to difference."

There are interesting glimpses in Lefebvre's spatial ideals about the body and corporeal sensuality of the Mexican poet, essayist, and Nobel Laureate Octavio Paz (1914–98). In *The Production of Space,* Lefebvre repeatedly draws on Paz's surrealist dialectical interpretations of the body and "signs of the body" (by means of mirrors) (cf. POS, p. 184; pp. 201–202; pp. 259–60). Lefebvre, too, uses an enigmatic Paz poem as the epigraph to *The Production of Space*. Meanwhile, he concurs with Paz's thesis of the "disjunction" of the body in Western Cartesian thought and its "conjunction" in the Eastern, non-Christian tradition. Imprisoned by the four walls of abstract space, our bodies are not ours, both Lefebvre and Paz remark; our sexuality gets refracted and mediated by mirrors of nonknowledge, by how we are *meant* to see ourselves in society. "Apart from the lack of fantasy and voluptuousness," Paz wrote, "there is also the debasement of the body in industrial society. Science has reduced it to a series of molecular and chemical combinations, capitalism to a utilitarian object—like any other that its industries produce. Bourgeois society has divided eroticism into three areas: a dangerous one, governed by a penal code; another for the department of health and social welfare; and a third for the entertainment industry."[14]

The right to difference cried out as loud as the right to the city. For Lefebvre, the two are commensurably united, tautologically woven into the fabric of any liberated space, any differential space, expressing a geography of "different rights," moving beyond simple "rights in general"—as Lefebvre put it in *Le manifeste différentialiste* (1970). The right to difference, he warned, "has difficulty acquiring a formal or judicial existence."[15] Indeed, rather than stipulating another "abstract" right among many, "it is the source of them."[16] If this program encroaches on the domain nowadays seen as "postmodern," Lefebvre preempts it as a humanist ideal, citing the German mystic Angelus Silesius (1624–77) for clarification: a flower doesn't reduce itself to one particular feature of nature; nature herself bestows particularity to a flower. A flower has its own specific form, its own smell, color, and vitality, yet it comprises the totality of nature, its cosmic universality, its *essential* powers.[17] "A rose is without a why," said Silesius, famously. "It flowers because it flowers." Thus, its very universality ensures its particularity, supports its discrete identity, just as, claims Lefebvre, Marx argued in *The Jewish Question* (1844) that human emancipation guaranteed political emancipation, rather than the other way around. Implementing the right to difference necessitates the "titanic combat between *homogenizing powers* and *differential capacities*. These homogenizing powers possess enormous means: models, apparatus, centralities, ideologies (productivism, unlimited growth). Such powers, destroying both particularity and differential possibility, enforce themselves through technicity and scienticity, and via certain forms of rationality."[18]

Differential capacities, on the other hand, often go on the defensive and usually can't express themselves offensively, as polycentric powers, united in heterogeneity against an abstract, homogeneous force—which spreads itself differently and unevenly across global space. The "titanic struggle" isn't straightforward; threats, Lefebvre recognizes, wait covertly in ambush, especially within the Marxist tradition, where the specter of Leninism, with its monolithic mentality, its doctrine of party and working-class universality, haunts the dialectic, shadows any "differentialist manifesto." Within space, this dilemma becomes at once simpler and more complicated. "The more carefully one examines space," Lefebvre explains, "considering it not only with the eyes, not only with the intellect, but also with all the senses, with the total body, the more clearly one becomes aware of the conflicts at work within it, conflicts which foster the explosion of abstract space and the production of a space that is *other*" (POS, p. 391). So, within abstract space, militancy foments within its lived interstices, within its lifeblood and organic cells:

> Thanks to the potential energies of a variety of groups capable of subverting homogeneous space for their own purposes, a theatricalized or dramatized space is liable to arise. Space is liable to be erotized and restored to ambiguity, to the common birthplace of needs and desires, by means of music, by means of differential systems and valorizations that overwhelm the strict localization of needs and desires in spaces specialized either physiologically (sexually) or socially. An unequal struggle, sometimes furious, sometimes more low-key, takes place between the Logos and the Anti-Logos, these terms being taken in their broadest possible sense—the sense in which Nietzsche used them. The Logos makes inventories, classifies, arranges: it cultivates knowledge and presses it into the service of power. Nietzsche's Grand Desire, by contrast, seeks to overcome divisions—divisions between work and product, between repetitive and differential, or needs and desires. (POS, pp. 391–92)

The Production of Space thereby underscores Nietzsche's contribution to the right to difference, to the prioritization of the lived over the conceived. Or, better, with Nietzsche (and Marx), Lefebvre seeks to transcend a factitious separation under modern capitalism, a compartmentalization between thinking and acting, between theory and practice, life and thought—dissociation and sundering that spelled alienation and *in*difference.[19] Lefebvre's attraction to Nietzsche here was highly personal and deeply political. The latter's insistence on overcoming the past and reaching out for the future, as well as the finger he gave to Christianity—expressed so vividly with the quip "God is dead"—had obvious appeal to somebody who'd seen his beloved sun crucified.[20] "They've crucified the sun! They've crucified the sun!" wailed young Lefebvre years earlier, resting under a giant crucifix during a long country walk in the Pyrenees. It was he who'd been crucified, he recounts in *La Somme et le Reste* (*Tome I,* pp. 251–52). Nietzsche showed Lefebvre how he could rescue the sun from the cross and, a little scarred, return the bright yellow ball to the sky where it belonged. At the same time, Nietzsche's critique of rationality, of universal truths and idols, of prime movers and systematized thinking spoke volumes to Lefebvre, who, like the other Marx, Groucho, struggled with any club that had him as a member.

In *The Birth of Tragedy* (1872), Nietzsche evoked the battle between Dionysian and Apollonian art forms. And though Lefebvre says this analysis is "inadequate," he nonetheless realizes that it's "certainly meaningful" with respect to "the dual aspect of the living being and its relationship to space" (POS, p. 178). Borrowing from Greek deities, Nietzsche said Dionysus and Apollo are two different cultural impulses, metaphors for our civilization and for our own personalities: the former favors irrational, unfettered creativity and self-destructive "paroxysms of intoxication"; the latter expresses rationality, harmony, and restraint, "the calm of the sculptor god." Lefebvre opts for Nietzsche's figure of Dionysus, walking a knife-edge path between coherent, ordered, dialectical logic (Logos) and irrational Dionysian spontaneity and creativity (Anti-Logos).

"Under the charm of the Dionysian," Nietzsche wrote, "not only is the union between man and man reaffirmed, but nature which has become alienated, hostile, or subjugated, celebrates once more her reconciliation with her lost son, man."[21] On the side of Logos, of Apollo, "is rationality, constantly asserting itself in the shape of organizational forms, structural aspects of industry, systems and efforts to systematize everything ... business and the state, institutions, the family, the 'establishment.' " On the side of Anti-Logos, of Dionysus, are forces seeking to reappropriate abstract space: "various forms of self-management or workers' control of territorial and industrial entities, communities and communes, elite groups striving to change life and to transcend political institutions and parties" (POS, p. 392).

With differential space, Lefebvre plays his Nietzschean–Marxist trump card at a decisive moment, as an innovative geographer whose ideals seem more akin to Orpheus than Prometheus. Marx's cult-hero was Prometheus, who suffered because he stole fire from the gods. It was he who appeared in the noble guise of the proletariat chained to capital. The Promethean principle is one of daring, inventiveness, and productivity, yet Lefebvre's Orphean spirit neither toiled nor commanded. It intervened unproductively, sang, partied, listened to music (to Schumann—his favorite), and reveled in a Dionysian space of drink and feast, of mockery and irony. Differential space isn't systematic, and so the form and content of *The Production of Space* unfolds *eruptively* and *disruptively,* unsystematically through a Nietzschean process of "self-abnegation." "I mistrust all systematizers," Nietzsche said; "I don't build a system," Lefebvre concurred, on the page and in politics. Nothing here even remotely resembles a system, the latter pointed out, neither in form nor in content. "It's all a question of living," he explained in closing lines of *Le manifeste différentialiste*. "Not just of thinking differently, but of *being* different," of uniting ourselves with our protean vital powers and constructing a spatial form worthy of those powers: a "true space," he labels it in *The Production of Space* (p. 397), "the truth about space."

* * *

In a 1985 preface to his earlier original, octogenarian Lefebvre filled in some gaps of an eleven-year-old thesis. Margaret Thatcher and Ronald Reagan had since stormed onto the scene, castigating an evil Empire and waging war in the South Atlantic. Meanwhile, the Berlin Wall tottered. Then it would topple, imploding from within while battered from without; an erstwhile absolute space, outside the realm of capitalist social relations, would shortly be colonized, rendered another abstract market niche. The production of space began edging itself outward onto the global plane, deepening preexisting productive capacity in traditional centers of power while pulverizing spaces elsewhere in the world, disintegrating and reintegrating them into a post-postwar spatial orbit. All hitherto accepted notions of national and local politics, replete with closed absolute frontiers, thus began to melt into air; a new fragmented, hierarchical, and homogeneous landscape—a "fractal" neocapitalist landscape—congealed.[22]

On a few occasions, Lefebvre brandishes the term *globality,* hinting at the continued planetary reach of this process, anticipating our own debates around globalization. Moreover, nobody could ignore, he said, the replacement of state-planning and demand-led economics by a "badly-reconstituted neo-liberalism," signaling not an end of planning per se but its reemphasis, a new machination of the liberal-bourgeois state, now unashamedly in cahoots with capital, notably with finance capital. This new state orthodoxy parallels the new production and control of global space, a "new world order," at once more rational and irrational in its everyday penetration and supranational subjugation.

During this same eleven-year period, a neoliberal right wing triumphed with its "metanarrative" of the market. Within the space of seventeen years, between *The Production of Space* and Lefebvre's death, in all walks of life—in politics and business, in business schools and universities, in peoples' imagination—a new plausibility about reality became common wisdom, dictating the terms of what is (and isn't) possible. Soon, all *oughts* were sealed off—like the Geneva headquarters of the World Trade Organization—behind a barbed wire fence of *is*. A global ruling class had set off on its long march, dispatching market missionaries here, spreading TINA (There Is No Alternative) doctrines there, cajoling and imposing its will of a constantly expanding world market, brooking no debate or dissent.

The state and economy steadily merged into an undistinguishable unity, managed by spin doctors, spin-doctored by managers. Abstract space started to paper over the whole world, turning scholars and intellectuals into *abstract labor* and turning university work environments into another *abstract space.* Suddenly, free expression and concrete mental labor—the creation and dissemination of critical ideas—increasingly came under assault from the same commodification Lefebvre was trying to demystify. Suddenly, and somehow, intellectual space—academic and ideational space in universities and on the page—had become yet another neocolony of capitalism, and scholars are at once the perpetrators and victims, colonizers and colonized, warders and inmates.

More and more, academic labor power is up for sale and there for hire. And their products—those endless articles and books—are evermore alienated, increasingly judged by performance principles, by publisher sales projections—or by their ability to *justify* the status quo. Thus, when writers and scholars enter the Lefebvrian fray, when they write about daily life and global space, they should think very carefully about whose daily life they're talking about, whose (and what) space they mean. When they write about radical intellectuals like Lefebvre, they should think about their own role as radical intellectuals, turning Lefebvrian criticism onto themselves, analyzing their own daily life and space at the same time as they analyze global capitalism. Better to bite the hand that feeds than remain a toothless intellectual hack, another cog within the general social division of labor.

Guts, as well as Lefebvre, are needed to resist the growing professionalization of ideas and university life, where, before all else, abstractions and cybernanthropes, evaluations and economic budgets sanction knowledge claims. Hence, a universal capitulation to the conceived over the lived hasn't just taken place in the world: it has taken place in those who should know better, in those who read Lefebvre's work, in those who edit and contribute to radical journals. When scholars write about emancipation, about reclaiming space for others, we might start by emancipating ourselves and reclaiming our own work space, giving a nod to disruption rather than cooptation, to real difference rather than cowering conformity. Yet before imagination can seize power, some imagination is needed: imagination to free our minds and our bodies, to liberate our ideas, and to reclaim our society as a lived project. That, it seems to me, is what the production of differential space is really all about. It's a project that can begin *this afternoon.*

7

GLOBALIZATION AND THE STATE

The space that homogenizes has nothing homogeneous about it.

—**Henri Lefebvre,** *The Production of Space*

Capitalist production has unified space, which is no more bound by exterior societies. … The accumulation of mass produced commodities for the abstract space of the market, just as it has smashed all regional and legal barriers, and all corporative restrictions of the Middle Ages that maintained the quality of artisanal production, has also destroyed the autonomy and quality of places. The power of homogenization is the heavy artillery that brought down all Chinese Walls.

—**Guy Debord,** *The Society of the Spectacle*

Henri Lefebvre never wrote anything explicit about what we today call "globalization." But his thesis on the production of space, and its role in the "survival of capitalism," tells us plenty about our spatially integrated yet economically fractured world. It's a world in which information technology collapses distances between continents and fosters cultural and market exchanges, yet simultaneously reifies "uneven development" between (and within) richer and poorer countries. It's a world, too, in which a lot of our economic, political, and ecological problems possess an intriguing geographical dimension, whose spatial stakes have ratcheted up a few notches since Lefebvre first unveiled his famous thesis: "Capitalism has found itself able to attenuate (if not resolve) its internal contradictions for a century, and consequently, in the hundred years since the writing of *Capital,* it has succeeded in achieving 'growth.' We cannot calculate at what price, but we do know the means: *by occupying space, by producing a space.*"[1]

Indeed, the question of space has now opened out to its broadest planetary frontiers. The economy expanding materially across the globe—and ideologically within daily life—has equally contorted the internal and external dynamics of the nation-state; "a new state-form," Lefebvre remarked, was asserting its will in the mid-1970s, a grip that has congealed into assorted superstate and suprastate authorities and agreements, replete with acronyms galore (e.g., WTO, GATT, NAFTA, FTAA, MAI, etc.). The planetary production of abstract space has detonated the traditional scale and scope of political management and warranted new contradictory modes of national and international (de)regulation—a new "State Mode of Production" (SMP)—opening up markets here, sealing them off there, lubricating free flows of capital on one hand, doling out subsidies on the other. And with the Berlin Wall gone, the former Communist Bloc can now be seduced, can now jump on the bandwagon and grab its market cachet—or be damned. The hoary demarcation between two rival systems—one with its abstract space of "freedom of choice" to purchase a dazzling array of consumer durables, the other with its absolute space of dictatorial personality and totalitarian rule—is no more. The rational combination of each rule has given liberal-bourgeois capitalism license to permeate all reality, to colonize all culture and dominate all geography. And, as we speak, that power of its market homogenization is quite literally poised to smash down all Chinese walls.

"No one," says Lefebvre, "would deny that relations between the economy and the state have changed during the course of the twentieth-century, notably during the past few decades." Enter the SMP, Lefebvre's attempt to shed light on this new general tendency, this new "qualitative transformation," "a moment in which the state takes charge of growth, whether directly or indirectly."[2] "The State Mode of Production" is the title of the third and most original volume of Lefebvre's four-tome exploration of the capitalist state, *De l'État,* penned furiously between 1976 and 1978 as fiscal crisis of the state raged at every level of government in advanced countries. In 1975, New York City declared itself fiscally bankrupt—President Gerald Ford told it famously to "Drop Dead!" In 1978–79, Britain underwent its "winter of discontent"; refuse and utility workers lobbied James Callaghan's Labour government for cost-of-living raises. Power cuts, garbage mountains, and rank-and-file acrimony greeted the prime minister's austerity appeals. And in Italy and West Germany, extraparliamentary volatility epitomized by the militant "Red Brigade" and "Baader-Meinhof" became the new disorder, filling the party political void, flourishing in the ruins of welfare-state Keynesian—capitalism with a human face—which was about to perish forever.

Lefebvre's theoretically dense quartet, drawing heavily on Hegel, Marx, and Lenin, wedges itself within this interregnum, when the Phoenix of "New Right" orthodoxy was set to rise out of Keynesian ashes. As is so typical with Lefebvre, much of this work is padded out with digressive and repetitive disquisitions on Mao and Stalin, on Lenin and Trotsky, on China and Yugoslavia, which have little or no resonance nowadays. On the other hand, equally typical are insights that are ahead of the game and live on: the new "materialization" of the state, at once a decentralization and reconcentration of governmental power and remit, signaled, Lefebvre reckoned, an epochal transition, a situation in which "the state now raises itself above society and penetrates it to its depths, all the way into everyday life and behavior."[3] Herein the SMP has several dimensions, and a few telling moments: a *managerial* moment of consent, a *protective* moment that seduces its population, and a *repressive* moment that kills, that monopolizes violence through military expenditure and strategies of war. Meanwhile, within the state apparatus resides a restructured "division of political labor," coordinated by technocrats, the military, and professional politicians, those agents of the state who preside over an abstract space that "at one and the same time quantified, homogenized and controlled—crumbled and broken—hierarchicized [*hiérarchisé*] in 'strata' that cover and mask social classes."[4]

Ironically, the Marxist clarion call of the "withering away of the state" in the passage toward socialism had been hijacked by an innovative and brazen right wing, while the Communist Left—Lefebvre's own constituency—bizarrely clung on to a statist crutch. The French Communist Party still insists on the importance of the state, Lefebvre said in an interview in 1976. "This is Hegelian thought; namely, the state is an unconditional political experience, an absolute. We cannot envisage neither its supranational extension nor its withering away, neither its regressive decomposition nor its regional fragmentation. To maintain the state as absolute is Stalinist, is to introduce into Marxism a fetishism of the state, the idea of the state as politically unconditional, total, absolute."[5] And yet the conservative flip side threatens society and economics, a "danger that menaces the modern world and against which it is necessary to struggle at all costs. There is no 'good state'; today there is no state that can avoid moving towards this logical outcome: the State Mode of Production; that's why the only criterion of democracy is the prevention of such an outcome."[6]

Lurking behind this new state form, behind a "simulacrum of decentralization," is thus a right-wing Hegelian "ruse of reason." The neoliberal state's divestment from the public sphere in the name of personal liberty—epitomized by Margaret Thatcher's 1980s maxim "There is no such thing as society, only individuals and families"—"merely transferred the problems," Lefebvre reckons, "but not the privileges."[7] No longer is government coughing up for public service provision and collective consumption budgets; instead it subsidized corporate enterprise, lubricated private investment into "the secondary circuit of capital," and left it to grassroots groups and voluntary organizations to clear up the mess of market failure, to handle affairs of redistributive justice.

The loosening or breaking down of the state's centralized administration, its apparent rolling back and strengthening of civil society, is really "the crushing of the social between the economic and the political."[8] Privatization and deregulation actually extend the domain of the state rather than restrict it. From being outside of civil society, the state henceforth suffuses all civil society. "If the state occupies three dominant sectors (energy, information technology, and links with national and world markets)," Lefebvre cautions, "it can loosen its reins somewhat towards subordinate units, regions and cities, as well as business ... it can control everything without needing to monitor everything."[9]

* * *

Abstract space and SMP orthodoxy have proliferated most forcefully in the post-1991 era. So forcefully, in fact, that the dialectical link between space and politics seems to have receded behind the blanket category of economic globalization. Neoliberal pundits like economist Richard O'Brien now suggest that because the economy is supposedly a totalizing force, footloose and fancy-free, everywhere and hence nowhere in particular, a Lefevbrian antithesis is in our midst: "The End of Geography."[10] Here big finance and mobile money arguably run roughshod over specific geographical contouring, like national jurisdictional boundaries, and trample over politics itself. So the "end of geography" is, for O'Brien, tantamount to "death of politics," the denouement of the New Right's withering away of the state, because there's now no political space for any alternative, no geographical niche or strategic spatial maneuvering for anything but neoliberal financial logic. It's an inexorable inundation that no Noah's ark can withstand. The SMP has calibrated society to such a finely tuned degree that it pervades everything and everybody. It's the economy, state, and civil society all rolled into one.

The post-Seattle Left has come up with its own, curious version of this thesis. In their Marxist blockbuster *Empire,* the hard-hitting duo Michael Hardt and Antonio Negri don't so much hail the end of geography as extend geopolitical frontiers to the absolute max. "Empire" is their slippery concept for disentangling a similarly slippery and entangled globalized world order, an order about as old as *The Production of Space*. Empire is different from the imperialist Empire of old, say Hardt and Negri; above all, it's the Empire of globalization, a new kind of "decentered" sovereignty, having no boundaries or limits—other than the limits of planet Earth. While it flourishes off U.S. constitutionalism and frontierism, Empire isn't simply American nor is the United States its center: Empire has no center. Its power dynamics don't operate like any Hobbesian *Leviathan;* power isn't repressive from the top-down, administered on the unruly rabble below. Rather, power is more "biopolitical," regulating people from *within,* seeping into subjectivity and through the whole fabric of society.[11]

The total control character of Empire diffuses through intricate "nonplace" and "deterritorialized" networks, which are tricky to pin down let alone resist. Such is very much an Althusserian geography of power, a geopolitical process without any clearly discernible subject or agent. Notwithstanding, Hardt and Negri welcome the advent of Empire, and they root for anything that will push it to its ultimate expanse; and the quicker the better! Here they're unashamedly Marxist in analytical scope, yet unequivocally *pro*globalization in their political hopes. Thus, despite its dread and foreboding, its abuses and misuses, "we insist on asserting that the construction of Empire is a step forward in order to do away with any nostalgia for the power structures that preceded it. … We claim that Empire is better in the same way that Marx insists that capital is better than the forms of society and modes of production that came before it" (p. 43). Within Empire are the seeds of its own demise: Empire, in short, produces its own grave diggers. The virtual world it commandeers can eventually become a "real virtuality," where a transnational working class achieves "global citizenship" (p. 361).

At that point, workers of the world will assert themselves as "the concrete universal," as "the multitude." Hardt and Negri deign for nothing less. The Left has to match a "deterritorialized" ruling class by inventing a "deterritorialized" politics of its own, tackling bad virtuality with good virtuality, fighting corporate globalization with civic globalization, confronting a fluid and faceless enemy on their terms, at the global scale. Here, the duo insists (p. 44), there's no place for "the localization of struggles." Now, within the global totality of capitalism, "place-based" activism is a bankrupted ploy: at best misconceived, at worst reactionary. "This leftist strategy of resistance to globalization and defense of locality is also damaging because in many cases what appear as local identities are not autonomous or self-determining but actually feed into and support the development of the capitalist imperialist machine." "It is better," Hardt and Negri conclude (p. 46), "both theoretically and practically to enter the terrain of Empire and confront its homogenizing and heterogenizing flows in all their complexity, grounding analysis in the power of the global multitude."

* * *

How, we might justifiably wonder, can resistance to global power begin if it isn't permitted to nurture somewhere, in a specific location? And what would be the point of *any* global politics if it isn't responsive to some place or people, isn't rooted in a particular context? Just as Marx in *Critique of the Gotha Program* (1875) accused Ferdinand Lassalle of "conceiving the workers' movement from the narrowest national standpoint," Hardt and Negri take it the other extreme, conceiving the workers' movement from the broadest international standpoint. In a document fundamental to Lefebvre's ideas on the state and politics, Marx critically assessed the draft program of the United Workers' Party of Germany, fronted by Lasselle: "It is altogether self-evident," Marx wrote, "that, to be able to fight at all, the working class must organize itself at home *as a class* and that its own country is the immediate arena of its struggle."[12] This class struggle, Marx added, must be national "in form" but not "in substance." The "substance" of the workers' movement, of course, is international. But Marx's internationalism retains dialectical content and real life *friction*. "To what does the German Workers' Party reduce its internationalism?" he queried. "To the consciousness that the result of its efforts will be '*the international brotherhood of peoples*.' Not a word, therefore, *about the international functions* of the German working class! And it is thus that it is to challenge its own bourgeoisie—which is already linked up in brotherhood against it with the bourgeois of all other countries."[13]

The key questions Marx posed in *Critique of the Gotha Program,* and that Lefebvre developed and extended in his work on the state—How can workers of the world conjoin across national contexts? Does civil society in itself have sufficient resources and organizational capacities to replace and reabsorb the state? How can *autogestion,* or workers' self-management, function *against* and *within* the state?—seem too mundane for Hardt and Negri's "Big Bang" thesis of global revolution. Lefebvre, needless to say takes a different tack; for him, "Marx's comments on the Gotha Program have lost nothing of their saltiness."[14]

"The ever-mounting number and size of government institutions in modern society," says Lefebvre, taking an inventory of the problem, "call attention as never before to the contradiction between the political and social aspects. … Will the modern state manage to stifle social life entirely under the crushing weight of politics? This is the question the Lassallians ignored, but that Marx never tired of raising." "What changes will the form of the state undergo in the new society?" "What social functions similar to the functions now performed by the state will remain in existence?" "In the transitional period, the objective is not simply to destroy the state (that is the anarchist position), but to let society as a whole—the transformed society—take over the functions previously performed by the state."[15] "Marx's objective," Lefebvre points out, "wasn't necessarily opposed to that of the anarchists: the end of the state, the end of hierarchies and political instances, with an attendant abolition of private ownership of the means of production."[16] Nonetheless, anarchists like Bakunin "abridge and even jump over the period of transition."

Hardt and Negri's political vision similarly abridges and leapfrogs the period of transition. For them space has gobbled up place and the global champed away at the local; ergo, the scale of politics has to be pitched at a still-unimaginable world space, with an "international brotherhood of peoples" ("the multitude") pitted against an omnipotent abstract state: "if we are consigned to the non-place of Empire," they write, "can we construct a powerful *non-place* and realize it concretely, as the terrain of a postmodern republicanism?" (p. 208). From this terrain, though, there's *no staging post for politics* and no grounding for struggle: the space of global politics cannibalizes the politics of place. Positing universality without particularity, the global without the national, negation without transition severs the dialectical mediation between form and content, between space and place, and cuts off bridge building between real people and real problems. *Empire,* consequently, parts company with Marx and Lefebvre's radical vision and leaves us nothing, in the here and now, to stave off death on credit. "The transformation of society," reasons Lefebvre, "defines itself first of all as an ensemble of reforms, going from agrarian to planetary reforms that imply the control of investment; but this sum of necessary reforms doesn't suffice: one needs to add to it something essential: the transformation of society is a series of reforms *plus* the elimination of the bourgeoisie as the controlling class of the means of production."[17]

* * *

Making space for a politics of place, and putting place in its reformable global space, is something Lefebvre's spatial dialectic does with remarkable prowess. The neocapitalist order, he recognizes in *The Production of Space* (POS in citations), has stripped space of its naturalness and uniqueness, giving a "relative" character to erstwhile "absolute spaces," transforming them into something more "abstract." Absolute space was "historical space," "fragments of nature," located on sites that were chosen for "intrinsic qualities" (POS, p. 48): caves, mountaintops, streams, rivers, springs, islands, and so forth. This was a natural space, Lefebvre says, "soon populated by political forces." Colonization was an early driving force behind such politicization; economic, administrative, and military organization got inscribed in parts of the world once unique, once outside centers of power and domination. At that moment, abstract space took over from historical space, setting in motion a new historical and geographical dynamic.

In its birth pangs, Marx called this impulse "primitive accumulation"; in *The Production of Space,* Lefebvre gives it an explicit spatial dynamic. "The forces of history smashed naturalness forever," he notes, "and upon its ruins established the space of accumulation (the accumulation of all wealth and resources: knowledge, technology, money, precious objects, works of art and symbols)" (POS, p. 49). What Hardt and Negri identify as "Empire" is, in reality, the most developed form yet of Lefebvrian abstract space, and it incarnates the passage from "the capitalist state" to the SMP, replete with its own biopower: "The state's management," he says in *De l'État,*

> develops its effects in society as a whole; it doesn't limit itself to steering society: it modifies society from top to bottom. Political society engenders social relations; reacting in the breast of civil society, political society modifies these social relations with a "determined" orientation: formation, consolidation and reinforcement of the middle-classes. This process can itself be considered as a political product, because its relations tend to reproduce themselves in assuming the general reproduction of social relations of production and domination. … The state redirects the reproduction of social relations by diverse means: by repression and hierarchy, by the production of appropriated (political) space, in brief, by the management of all aspects of society.[18]

The domain of Empire thereby periodizes a sort of neoabstract space, something even more abstract than heretofore, whose generative roots hark back to the global crises of the mid-1970s, to economic and political upheavals triggered by the demise of Bretton Woods and catalyzed by the 1973 oil embargo. Yet there's a crucial difference between Lefebvre's theorization of abstract space and SMP and Hardt and Negri's Empire: the former anchors a shifting capitalist reality solidly in sidewalk space rather than cyberspace. Instead of becoming groundless, free-floating, and virtual, abstract space, under a super- and suprastate mode of production, exists materially in absolute space: the latter may have been replaced, but it certainly hasn't *disappeared.* Absolute space survives, lives on, Lefebvre stresses, as "the bedrock" of abstract space; the distinction is historical not ontological: abstract space is meaningless outside of absolute space, outside of some physicality, much as the market price for diamonds doesn't make sense without a diamond mine. Marx reaffirms this message: "it appears paradoxical," he said in *Capital,* "to assert that uncaught fish are means of production in the fishing industry. But hitherto no one has discovered the art of catching fish in waters that contain none."[19]

The transition from absolute to abstract space mirrors the transition Marx identified between concrete and abstract labor: "productive activity (concrete labor) became no longer one with the process of reproduction which perpetuated social life; but, in becoming independent of that process, labor fell prey to abstraction, whence abstract social labor—and abstract space" (POS, p. 49). All labor, of course, is expenditure of human energy in a particular form, with a definite aim. In its differentiated quality, as "concrete labor," it creates use values; in its capitalistic guise, concrete labor assumes an undifferentiated character, becomes objectified in a commodity, and enters the world as exchange value. Then what was once the specific activity of, say, making a shirt immediately becomes general, becomes labor that's conditioned by the socially necessary labor time required to make shirts. "In its first aspect," wrote Marx, "labor presents itself as a given *use-value* of the commodity; in its second, it appears as money, either as money proper or as a mere calculation of the price of a commodity. In the first case we are concerned exclusively with the *quality,* in the second with the *quantity* of labor."[20] This second kind of labor, labor toted up as value and exchange value, is what Marx tagged "abstract labor."

Marx makes an analytical distinction rather than a real-life separation and shows Lefebvre how to keep the link between the specific and the general, quality and quantity, use value and exchange value, and the concrete and the abstract in taut dialectical tension. You can't have one without its "other": without use values commodities would have no exchange values; if they don't have exchange values, tailors and shirt manufacturers, at least in our society, would stop making shirts. Similarly, abstract space is reality only insofar as it is embedded in absolute space; space has reality only insofar as it is embedded in place. Absolute space lives on as a basic empirical building block, as the ontological layering of society, just as the dynamics of daily life still respond to classical Newtonian physics—even after Einstein's revolution, even after Heisenberg's "uncertainty principle." Atomic particles may be in two places at once but people can't; daily life—lived space of representation—is always couched in absolute time and space, is always located and locatable in observable empirical reality. Place matters for life and for politics. And it poses dilemmas for theory.

Lefebvre's spatial triad tackles the theoretical conundrum by giving a dialectical contradiction a "triple determination." Absolute and abstract spaces, for him, become two different guises of a unity straddling three identifiable moments: the conceived–perceived–lived. Here it's possible to recognize how conceived spaces of representation are geared toward the production of abstract space, with its global reach, while absolute space is the locus of perceived spatial practices and daily life. The process world of abstract space is the representation of space commandeered by the rich and powerful, by CEOs and cybernanthropes, by state politicians and free-market planners, by those who're invited to global summits like the World Economic Forum, who conceive spaces every year at

Davos, Switzerland. Representations of space are likewise projected onto lived reality by the World Trade Organization and the World Bank, and conjured up on endless corporate and state flip charts in neat boardrooms and cabinet offices, cordoned off from the messy disorder of lived experience outside. The reality of these abstract representations provide the context for neoliberal spatial practices across the globe, as well as the grassroots activism scattered around assorted spaces of representation.

Neither realm—the absolute lived or the abstract conceived—exists as opposites in a binary; yet neither are they the same reality. They are, in fact, two instances within one world, relative distinctions within a unity, definite relations within neocapitalist globality. Lefebvre is smart enough to know not only the global forest and its constitutive trees; he also knows how each realm is *mediated.* He knows how mediation resides *within,* not between, each moment: mediation isn't a third piece to slot into a gap. Lefebvre knows the mediation between space and place, between the abstract and the concrete is intrinsic to each respective opposite. Spatial practices, those practical routes, networks, and received actions ingrained and normalized within lived experience, play a crucial mediating role in global space. They keep the global and the local scales together, yet apart.

On the one hand, everyday spatial practices make the local seem absolutely local; on the other hand, they make the global, especially as it is filtered through the TV or chronicled in the "International News" pages of the dailies, seem absolutely global, as something beyond the reach of any place-bound locality. Spatial practices thus pivot around the "thing" world of everyday life, and they internalize both representations of space and spaces of representation. But this everyday perceptual thing-world is flush with processes and representations that aren't graspable from the level of perception and lived experience alone—rather like abstract labor isn't graspable from the standpoint of concrete labor. Consequently, spatial practices are in the thrall of conceived space, yet they have the latent capacity, Lefebvre says, to subvert the conceived, to "detonate" lived space and to transform the global—but only if the severed can be reconnected and separated commingled.

* * *

Lefebvre has no truck with binary thought and sundered practice. Hence he has sound reasons for positing a triad. To begin with, he wants to ensure that space doesn't simply get equated to the abstract and place doesn't get equated to the concrete. But, neither, too, does he want to give credence to the opposite view. He doesn't accept that space has overwhelmed place—that in our high-tech, media-saturated society space has decoupled from its place mooring. The idea that reality is now rootless and "nonplace" would strike Lefebvre, the grand theorist of everyday life, as patently ridiculous and politically dubious. He would thereby rally against the "network society" promulgated by former colleague Manuel Castells; namely, the "space of flows" has substituted "the place of spaces." This ontological binary is something Hardt and Negri's epistemology revels in, with its either–or mentality: "In this smooth space of Empire," they say, "there is no *place* of power—it is both everywhere and nowhere. Empire is an *ou-topia,* or really a *non-place* … abstract labor is [now] an activity without place … exploitation and domination constitute a general non-place on the imperial terrain."[21] And, to redouble the point, they add (p. 237), "Having achieved the global level, capitalist development is faced directly with the multitude, *without mediation.*"[22]

Contra Hardt and Negri, Lefebvre says, "Everything weighs down on the lower 'micro' level, on the local and localizable—in short, on the sphere of everyday life" (POS, p. 366). Indeed, everything—the global included—"*depends* on this level:

exploitation and domination, protection and—inseparably—repression" (POS, p. 366; emphasis in original). Why else would he insist that everyday life "is the inevitable starting point for the realization of the possible?"[23] All of which "doesn't mean that the 'micro' level is any less significant" or necessarily reactive or reactionary. "Although it may not supply the theater of conflict or the sphere in which contending forces are deployed, [the local] does contain both the resources and stakes at issue" (POS, p. 366). Here place assumes the "form," if not the "substance," of grassroots struggle, which always unfolds willy-nilly within a national cultural and legal context: Marx emphasized as much.

That the global appears omnipotent in daily life, that its corresponding ideology insists globalization is somehow inevitable or natural, is, for Lefebvre, only to reinforce a premise of "dissociation and separation." Dissociation and separation "are inevitable in that they are the outcome of a history, of the history of accumulation; but they are fatal as soon as they are maintained in this way, because they keep the moments and elements of social practice away from one another" (POS, p. 366). What so completely shatters and submerges the everyday, Lefebvre warns, is the active subversion—in theory and in practice—of all that constitutes the everyday: its separation from what is supposedly "non-everyday."

Abstract space, we might say, is everyday or it's nothing at all: absolute space always offers an everyday entry point for confronting the global sway of abstract space. Lefebvre insists on this vital fact, without which grassroots leverage would be neither possible nor permissible. Moreover, recent history seems to endorse Lefebvre's line, because absolute qualities of place have actually prompted progressive defiance *within* abstract space: they've prized open modest little holes within the global capitalist fabric, established noteworthy nodes of resistance within the capillaries of abstract power. In August 1999, for instance, a few months before the ruling classes got sleepless in Seattle, three hundred men, women, and children in Millau, southwest France, tore apart a partly constructed McDonald's eatery. This was an "organized dismantling," a Lefebvrian moment, spontaneous yet carefully planned by ewe's milk farmer José Bové, longtime rural militant and unionist, and cofounder of the *Confédération Paysanne*. In the wake of this widely reported event, Bové was arrested and accused of a million dollars worth of criminal vandalism. Bail money poured in fast, particularly from U.S. farmers, and an erstwhile working-class Frenchman became a radical folk hero around the world.

Specifics date back to February 1998, when the World Trade Organization, responding to the American meat lobby, condemned the European Union's refusal to import American hormone-treated beef. The proneoliberal World Trade Organization gave Europe fifteen months to open its frontiers, or else. The deadline expired in May 1999, and almost immediately the Clinton administration responded, slapping 100 percent customs import surcharge on assorted European products, Roquefort cheese included. Overnight, the price of Roquefort doubled from $30 a kilo to $60, effectively prohibiting sale. "We sell 440 tonnes of cheese annually to the states," Bové explained in his aptly titled book *The World Is Not for Sale*, "worth 30 million francs [€4.5 million]. Given that the cost of [ewe's milk] is half the value of the Roquefort, the producers are losing 15 million francs [€2.2 million]."[24] Ewe's milk farming is crucial to the economy of the Larzac region and a livelihood for Bové. "So on the one side of the Atlantic a wholesome product like Roquefort was being surcharged, while on this side we were being forced to eat hormone-treated beef."[25]

Lefebvre lists among absolute space such items as caves, mountaintops, springs, and rivers (cf. POS, p. 48). Roquefort cheese is "absolutely" unique in its production process; it's a special alchemy of ewe's milk, bread, and the caves of Aveyron's Massif Central, whose natural damp and airy grottos spawn "penicillin roqueforti," the mold that gives the cheese its distinctive blue veins. On the other hand, McDonald's signifies *malbouffe*—junk food—"food from nowhere," Bové calls it, and pits big multinationals against low-paid agricultural workers and small farmers in a complexly mediated class war. McDonald's symbolizes "anonymous globalization," having little relevance to real food or to local cultures; abstract food (and drink) thereby helps produce and reproduce abstract space.[26] In Millau, Bové's protest was a localized action that enabled people to see an abstract enemy more definitively, helping them to grasp concretely new forms of alienation and economic domination. It also ignited debates about globalization in France and elsewhere in the world; questions that Lefebvre posed theoretically decades earlier have been—and continue to be—explored on the streets of hundreds of cities everywhere.

Henceforth a new kind of multinational trade unionism in the agricultural sector has burst forth, politicizing urban streets, opening out onto a global stage, "one that denounces inequalities and struggles for work and for a redistribution of public funds, and has an international outlook."[27] Add this to the struggle for immigrant rights, for *sans papiers,* for the excluded and the homeless, together with local negotiations with the French state on a thirty-five-hour week—which Bové sees as all part of the movement against neoliberalism—then "a general upsurge is in the air." Meanwhile, Bové and the *Confédération Paysanne* participate in a bigger international umbrella movement, *Via Campesina,* which organizes and coordinates assorted farmers' associations and peasants' groups throughout North, Central, and South America; Asia; Africa; and Europe. *Via Campesina* mobilizes member organizations in their respective places, where they retain a fierce loyalty to local culture and local food systems, but their political activism bonds with other people elsewhere, reaches out across abstract space; in the contact zones a robust, mediated concrete politics takes hold.

What transpires here isn't so much the power of a homogeneous "multitude" as a "differentialist manifesto," a unity of diverse peoples, the potentiality of embodied radicalism, a dogged means toward some bigger global end. The Millau protest was an ostensible trivial "moment" within a circumscribed lived experience. And yet, it made the abstract space of globalization—with its intensification of agricultural production, genetically modified food, low-wage economics, international tariff and trade agreements—real and down-to-earth, part of everyday life. These phenomena already had been everyday, but they're often difficult to spot at ground zero. Some form of militancy is required to tease them out, to point a finger, to make people think and act. The Lefebvrian moment, in theory and practice, can still perceive a possibility—a crack in the edifice—help people name in everyday life a remote process, and make it palpable and hence challengeable and changeable. Lefebvre bequeathed us the theoretical "thoughtware" to give *body* to an abstraction, to highlight latent political ambitions before the means necessary to realize them have been created.

"Pressure from below," he stated near the end of *The Production of Space* (p. 383), "must confront the state in its role as organizer of space. ... This state defends class interests while simultaneously setting itself above society as a whole, and its ability to intervene in space can and must be turned back against it, by grassroots opposition, in the form of counter-plans and counter-projects designed to thwart strategies, plans and programs imposed from above." Challenging centrality inevitably behooves pluralism, an assault to central power by diverse local powers, by militant actions linked with specific grievances in specific territories.

Just as inevitable will be the centralized state's attempt to isolate or exploit the weaknesses of these local upsurges, of any place-based activism. "Hence a quite specific dialectical process is set in train" (POS, p. 382): "on the one hand, the state's reinforcement is followed by a weakening, even a breaking-up or withering away; on the other hand, local powers assert themselves vigorously, then lose their nerve and fall back. ... What form might resolution take?" What kind of local activism can ground itself in a global spatial practice? How can people find the means to recognize that their particular grievances coexist with other particular grievances elsewhere, and that these particulars need each other if they are to grow universally strong? How can people bond with other people across space and time, generalize as they particularize?

For Lefebvre, there is one path and practice that can bring people together to oppose the "omnipotence" of the state and multinational capital. This, he says, will be the "form taken today by the spontaneous revolutionary; it is no more anarcho-syndicalism" but something radically different, something even radically different from what Marx ever imagined: *autogestion*—democratic participation, workers' self-management, and control of ordinary peoples' destinies.[28] This isn't a revolutionary "recipe," Lefebvre cautions: autogestion must be perpetually negotiated and enacted, relentlessly practiced and *earned.*

Autogestion is never a condition once established but a struggle continually waged; it's a progressive strategy, not a consolidated model. Autogestion, he says, "carries within itself, along with the withering away of the state, the decline of the Party as a centralized institution monopolizing decision-making."[29] Autogestion is antistatist: it tries to strengthen the "associative ties" in civil society. Workers' militancy, in and beyond established trade unions, embodies autogestion, where rank and filers assert control of the means of production not simply comanage them with capital. Direct action, like the dismantling of McDonald's, living wage campaigns, workday reductions, urban housing squabbles, environmental justice, peasant land struggles, and the activities of *Via Campesina, Conférération Paysanne,* Global Exchange, Direct Action Network, Reclaim the Streets, the World Social Forum—

the list is endless and scattered everywhere across the globe—all hint at potential routes and pathways of autogestion, of transforming everyday life, of avoiding a spaced-out global ambition. "For me," Lefebvre said, back in 1966, "the problem of *autogestion* shifts more and more away from enterprises towards the organization of space."[30]

Despite its spontaneous genesis, autogestion will nonetheless unfold over the long haul, by hook or by crook, steadily and stealthily, pragmatically and politically. Everyday life cannot be transcended in one leap:

> But the dissociations that maintain the everyday as the "down-to-earth" foundation of society can be surmounted in and through a process of *autogestion*. Attentive and detailed study of the May 1968 events may yet produce surprises. There were tentative, uneven attempts at *autogestion,* going beyond the instructions that the specialized apparatuses handed down. … *Autogestion* points the way to the transformation of everyday life. The meaning of the revolutionary process is to "change life." But life cannot be changed by magic or by a poetic act, as the surrealists believed. Speech freed from its servitude plays a necessary part, but it is not enough. The transformation of everyday life must also pass through institutions. Everything must be said: but it is not enough to speak, and still less to write.[31]

* * *

" 'Think globally, act locally,' is still one of the best slogans progressives ever slapped across the backside of a vehicle," said *The Nation* in an editorial not too long ago (February 18, 2002). The virtues of a micropolitics of everyday life, of a "grassroots globalism," was made emphatic as a post–9/11 New York City was abuzz with thousands of demonstrators rallying against the World Economic Forum's session there. There's every reason to get excited by these demos against the World Economic Forum and the parallel good-guy gatherings at the World Social Forum in Porto Allegre, Brazil, and the possibilities they present for autogestion. But activists in the United States can, *The Nation* also reiterated, bolster this movement if they pressure their local representatives who continually vote the corporate line in Congress. One of the biggest globalization fights of 2001 was decided by a single vote when the House went 215 to 214 in favor of George W.'s fast-track authority for the Free Trade Area of the Americas treaty. This newfangled beefed-up NAFTA, destined to spin millions of people into a spidery neoliberal web, could have been nipped in the bud: only two votes were necessary to deny Bush legislative power to proceed. So as militants prepare for the next Seattle, they should at the same time turn up the hometown heat, create a little ruckus in everyday life, in their own as well as in their local politician's. Empire may or may not be the new formidable form of world governance and domination; but it's at home, outside one's own front door, in absolute space, where most of us can find our little place in abstract global politics.

8

Mystified Consciousness

> To have slaves is nothing; but what is intolerable is to have slaves and to call them citizens.
>
> —**Denis Diderot,** *L'encyclopédie*

At the beginning of 2005, a headline in the *New York Times* grabbed my attention: "Lessons, from Hitler's Germany, on the Perils of Religion" (January 15, 2005). The article recounted a speech given to a startled audience at the Leo Baeck Institute by historian Fritz Stern, professor emeritus at Columbia University. Stern warned of the dangers posed to the United States (and the world) by the rise of the Christian Right and reminded listeners that "Hitler saw himself as the instrument of providence." In the 1930s, said Stern, "some people recognized the moral perils of mixing religion and politics, but many more were seduced by it. It was the pseudo-religious transfiguration of politics that largely ensured his success."

Coming hot on the heels of Bush's second-term election victory, the admonition understandably raised a few eyebrows. Despite blue-chip tax dodging and squandering public monies, doctoring election ballots the first time around and fabricating a need for war, George W. seduced Christian conservatives and Bible Belt bigotry to sweep to power. "The comparison between the propagandistic manipulation and the uses of Christianity, then and now, is," cautioned Stern, "hidden in plain sight. No one will talk about it. No one wants to look at it."

Around the time of the appearance of this article, I was reading Henri Lefebvre's *La Conscience Mystifiée,* a book seemingly forgotten and largely ignored in Lefebvre's œuvre, a text still not translated into English. Moreover, few Anglophone critics even allude to what may be his most relevant political tract, seventy years after its original publication. Stern's concern about propagandistic manipulation and uses of Christianity, and the seduction of an electorate, was precisely Lefebvre's concern. I was as guilty as anybody for overlooking Lefebvre's attempt to comprehend such "mystified consciousness," and so was finally getting down to studying his initial claim to fame, a thesis published in 1936 as the Popular Front stormed to victory in Spain and socialists won out in France. Just as few American liberals, against a backdrop of failed war and economic mismanagement, could have predicted a romping neocon success, the Popular Front—uniting socialists, communists, and fellow-traveling lefties—believed its mandate would be a beachhead against Hitlerism. Little did people know what lay ahead.

In France, Lefebvre's book (written in collaboration with Norbert Guterman) was frowned on in the inner circles of the French Communist Party. Some of its contents seemed directed more at old friends than at new enemies. Pride was piqued; loyalties were tested. Workers were critiqued; classical Marxist tenets impugned. Despite communist wishful thinking, proletarian class consciousness isn't an "objective" category, Lefebvre said, isn't something singular and pure, absolutely distinct from bourgeois consciousness. Indeed, the German situation revealed the mismatch between "economic consciousness" and "political consciousness," given German workers voted—as their American counterparts had—against their own "objective" interests. Old friends began disowning Lefebvre; George Politzer, defending the mind of the proletarian masses, told his former comrade, in no uncertain terms, "There is no mystified consciousness; there are only those who mystify."[1]

And yet, Hitler marched on, and the people cheered and followed, carried along by the tidal wave of mass adoration. With eight million workers without jobs, Hitler promised work, promised bread and circuses. We will make arms again, he proclaimed, we will get our factories moving again. Workers cheered even louder, even while they gave themselves over—as both cannon fodder and economic fodder—to big financiers and monopoly capitalists, with petit bourgeois merchants in tow. "The bourgeoisie," Lefebvre said in *La Conscience Mystifiée,* in a second-term tongue, "is obliged to maneuver great masses of men awakened to a certain social and political consciousness. The bourgeoisie doesn't need ideas too refined and metaphysical. Carefully instigated banalities are usually more useful than metaphysics. It now needs only to utilize old everyday sentiments, sentiments whose fragrance is 'all natural' and 'simply itself': faith, hearth, race, heroism, purity, duty—banalities inscribed in all our hearts."[2]

Demagogues wax lyrical in simple yet seductive language; they instigate festivals and launch wars, create external enemies apparently more fearful than internal enemies: "they tenderly embrace infants, or eat soup with unemployed workers and soldiers; they ennoble work, and arouse sacred emotions. Amplified by a servile press, these shameful machinations glory in heroism."[3] With mystified consciousness, fiction just as easily transmogrifies into fact;

the age of reason reverts to a dark age of witches and witch hunts, of primitive barbarism and axes of evil.

* * *

Lefebvre made a long trip to Germany in 1932, and there's a breeziness and carefree air to his descriptions in *La Somme et le Reste,* of taking solitary country walks, hopping from one youth hostel to another, swigging beer with young communists in rustic inns, swaying to melodies of the *Threepenny Opera.*[4] Lefebvre spoke to many different people, observed the German situation firsthand, gasped at "the exalted ardor of the Hitler Youth," which clashed with the "enormous rigidity of the German Communist Party and its administrative apparatus," whose "brutal internationalism" was almost as brutal as Hitler's nationalism.[5] Lefebvre felt the enormous power of a volcano about to blow, as economic crisis deepened and unemployment grew. The desperate plight of workers seemed to presage political revolt, probably growing support for the communists.

He returned to France anxious about Hitler yet optimistic that misery and "unhappy consciousness" would prompt German working classes to do the right thing. International communists everywhere thought Hitler was a passing phase, something destined to fizzle out, a little like U.S. progressives thought Bush was a passing phase. However, in 1933, forty million Germans voted for Hitler, many communists included. Rank and filers had been urged by the apparatchik to rid the government of the liberal bourgeoisie. The German Communist Party hated the Social Democrats even more than they hated the National-Socialists, and in their zeal to overthrow capitalism they'd given a green light to fascism. Proletarians thereby displaced their angst rightward, not leftward, and the party hardly set an example. Workers acted counter to the classical communist texts: once, they had nothing to lose but their chains; now, they enchained themselves, reentered the cave to stare at shadows, and had been betrayed by the party. They'd been seduced and manipulated by the National-Socialists and let themselves be duped. "One needed to explain this fact theoretically," Lefebvre said.[6]

Several factors shaped Lefebvre's Hegelian–Marxist leanings, leanings that would stake out the theoretical coordinates of *La Conscience Mystifiée*. For a start, he'd been impressed by a series of pathbreaking essays written in 1926 and 1927 by philosopher–theologian Jean Wahl (1888–1974), exegeses destined to figure in an influential book titled *Le Malheur de la Conscience dans la Philosophie de Hegel* (1929). The buzzword here is *malheur de la conscience*—unhappy consciousness—which Wahl gleaned from Hegel's *Phenomenology of Spirit*. Hegel said all philosophy, thought, and history hinged on a "dialectical movement," where categories of the mind and reality exist in "immanent unity." Hegelian history is an immense epic of the mind striving for unity, attempting to free itself from itself. From a starting point purified of every empirical presupposition, Hegel's *Phenomenology* generates the objective world as a wholly internal movement of the mind, the mind constantly overcoming itself in a series of theses, antitheses, and syntheses. "Consciousness itself," Hegel said, "is the *absolute dialectical unrest,* this medley of sensuous and intellectual representations whose differences coincide, and whose identity is equally dissolved again."[7]

Within that restless history, "unhappy consciousness" struck like a pathological version of what Freud would later term "normal unhappiness." The inability of consciousness to reconcile itself, in both its particular and universal forms—to be itself as subject and "other" as object—is the source of great inward disruption in people. "Thus," claimed Hegel, "we have here dualizing of self-consciousness within itself, which lies essentially in the notion of mind; but the unity of the two elements is not yet present. Hence the Unhappy Consciousness. The Alienated Soul is the consciousness of self as a divided nature, a doubled and merely contradictory being." Hegel thought unhappy consciousness is like gazing at one's own self-consciousness in somebody else's consciousness. Consciousness was real yet somehow out there, elsewhere, unable to understand either its own thinking or the conditions that surround it. People exist in a mist-enveloped world, cut off from themselves and other people. In such a context, thinking "is no more than the discordant clang of ringing bells," said Hegel, "or a cloud of warm incense. … This boundless pure inward feeling comes to have indeed its object; but this object does not make its appearance in conceptual form, and therefore comes on the scene as something external and foreign."[8]

Therein, reckoned Wahl, lay the pervasiveness of alienation and the tragedy of human history. Wahl wasn't interested in the formalism of the Hegelian dialectic or in the "master–slave" contradiction Alexandre Kojève illuminated a decade on; instead, Hegel's emotional and spiritual content shone through. In Wahl's eyes, Hegel was an antecedent of Kierkegaard and a kindred spirit of Pascal; Hegel's dialectic, Wahl believed, was first and foremost intuitive and experiential, not conceptual and intellectual, something felt rather than thought. "The dialectic," Wahl wrote, "before being a method, is an experience by which Hegel passes from one idea to another. … It is, in part, a reflection of Christian thought, of the idea of a God made man, which led Hegel to a conception of the concrete universal. Behind the philosopher, we discover the theologian, and behind the rationalist, the romantic. … At the root of this doctrine, which presents itself as a chain of concepts, there is a sort of affective warmth."[9]

Lefebvre recognized how the abstract, idealist basis of Hegel's *Phenomenology of Spirit* could be made more materialist—indeed *should* be grounded in concrete history, in grubby actuality. Before long, he'd put a distinctively *political* spin on Wahl's religious interpretation of Hegel, stressing the social and structural origins of "unhappy consciousness," calling for solutions rooted in *praxis* not faith. This trajectory was immeasurably aided by another big formative event for Lefebvre: the rediscovery of the Hegelian origins of Marxism, as evinced with the debut appearance in the early 1930s of Marx's *Paris Manuscripts of 1844*. Almost immediately he and Norbert Guterman began translating and popularizing Marx's early writings, extracting them alongside snippets from Hegel.[10] The duo would drag Hegel closer and closer to Marx and Marx closer and closer to Hegel. In the process, they'd stake out a rich, heterodox Hegelian–Marxism in between. "The importance of Hegel's *Phenomenology*," wrote the young Marx, in a text many times studied by Lefebvre, "and its final result—the dialectic of negativity as the moving and producing principle—lies in the fact that Hegel conceives the self-creation of man as a process, [and] objectification as loss of object, as alienation."[11]

Yet the Hegelian fix to alienation, to unhappy consciousness, had recourse to the thinking alone, to abstract dialectical logic. All the drama is in the head; everything is form without any real content, any real materiality and concrete objectivity. People populate Hegel's universe, but, as Marx said, they course around as mere "forms of consciousness," as minds without men. Hegel, for Marx, "turns man into the man of consciousness, instead of turning consciousness into the consciousness of real men." As such, Hegel posited a philosophical history not a philosophy of history. "This movement of history," Marx claimed, "is not yet the *real* history of man as a given subject."[12]

Lefebvre's originality lies in how he unites Hegel's unhappy consciousness with young Marx's humanist critique. In the mix, he warned neither individual nor collective forms of consciousness necessarily represent a criterion of truth: modern consciousness is a consciousness manipulated by ideology, ideology propagandized by state power. Different kinds of authority enter into people's heads, fill their minds, and mist their brains. *La Conscience Mystifiée* shattered the prevailing Marxist idea that working-class consciousness had transparent access to reality, that it somehow reflected in its collective head what was really out there. The thinly disguised target here was comrade Georg Lukács, whose influential *History and Class Consciousness* (1923) became a staple for Third Internationalist Marxists.[13] Crucial for Lukács had been the concept of "reification"—of how, in capitalist society, relations between people take on a "phantom objectivity" as relations between "things." Lukács located this "thingification" in the "commodity-structure" and suggested it was no accident that Marx began *Capital* with an analysis of commodities. The "fetish" commodity, said Lukács, became decisive "for the subjugation of men's consciousness … and for their attempts to comprehend the process or to rebel against its disastrous effects and liberate themselves from servitude of the 'second nature' so created."[14]

Marx's concept of fetishism illustrated how the human world becomes a fuzzy reality concealed by a material object, by a thing exchanged for another thing, money for money, money for labor power.[15] People, as labor power, as peculiar commodities, are separated from their activity, from the product of that activity, from their fellow workers, and ultimately from themselves. Isolation, fragmentation, and reification ensue. Needless to say, the ruling class prospers from all this, while the proletariat becomes submissive, unable to grasp their real conditions of life. Lukács reckoned reification could be punctured, exposed by the knowing mind acting on full knowledge of itself, acting in a "unified manner," understanding the "totality of history." Thus, in the climax to "Reification and the Consciousness of the Proletariat," the central pillar of *History and Class Consciousness,* Lukács concluded, "Reification is, then, the necessary immediate reality of every living person in capitalist society. It can be overcome by constant and constantly renewed efforts to disrupt the reified structure of existence ... by becoming conscious of the immanent meanings of these contradictions for the total development ... only then will the proletariat become the identical subject-object of history whose praxis will change reality."[16]

For Lefebvre, mystification stems from reification and is related to fetishism, but it is something qualitatively different as well, a realm deeper and more complex than reductionist economics and Marxist philosophy posited as a "mirror of knowledge." With mystification, the veil that's thrown over economic life is thrown over politics and psychology, over voting patterns and the totality of daily life. Lefebvre didn't buy Lukács's notion that the proletariat is "the identical subject-object of history," and he didn't see Marxism and dialectics as "universal."[17] Instead, Lefebvre's Marxist line asked, Given proletarian foibles and mystification, how can "a true and revolutionary consciousness be created"?[18] Dialectical thought and analysis can help, though only with an openness, flexibility, and honesty about "actual forms of alienation."[19] Mystification is so much more difficult to access when there appears no more to be mystification.

* * *

In *La Conscience Mystifiée,* Lefebvre flags out two forms of consciousness: *la conscience du forum* and *la conscience privée.* The first is a public consciousness, something social and collective; the second is individual and more bourgeois. Since the French Revolution of 1789, and the birth of modern capitalism, these two realms have increasingly split apart; modern Western philosophy has perpetuated the separation between society and the self, between the collective and the individual, between public life and private life. For Lefebvre, public consciousness tends to be progressive, pushes toward a higher abstraction, toward a reason potentially more liberating, more socialist. Individual consciousness is more retrogressive, likely to be impregnated with a theological ideology, usually with Christianity.[20] In our own times, this separation manifests itself as a glaring contradiction, as both a plague on the public realm and a warped, denigrated notion of individuality—what Lefebvre termed "an individualism against the individual." "The person of today," he says, "understands less than ever, spontaneously and immediately, their relation with society and its productive forces; instead of dominating this rapport, they are dominated by it, are victims of unconscious economic and social forces. One of the accomplishments of the dialectic has been precisely to find a theoretical unity between the private consciousness and the social consciousness."[21]

Private consciousness, for Lefebvre, is a consciousness *deprived*.[22] Here he plays on the Latin-rooted French word *privé,* with its dual meaning of "private" and "to lack," "to be deprived of." Thus, when people claim property as their own, "as mine," as privately controlled, the etymology of the term indicates this once constituted a loss, an act that deprived a larger public. The act of individual possession, the shibboleth of bourgeois society, likewise deprives the self of real selfhood. As Marx wrote in the *Economic and Philosophical Manuscripts,* "Private property has made us so stupid and one-sided that an object is only *ours* when we have it, when it exists for us as capital or when we directly possess, eat, drink, wear, inhabit it. … Therefore *all* the physical and intellectual senses have been replaced by the simple estrangement of *all* the senses—the sense of *having*."[23] The sense of "having," said Marx, is the only sense that really matters under capitalism. As a collective impulse, the desire to have and to possess—to organize a public consciousness around possession—drove a wedge between our subjective selves and our objective environment, between our private consciousness and our social consciousness.

Lefebvre pushes this logic further than Marx. On one hand, he warns of how public consciousness, the notion of collective will, can fall prey to ideologues and fascists who speak for "us," for a community based around nationhood and patriotism. In this regard, he knows how a false and forced unity produces what Nietzsche called a "herd mentality," a glorification of mediocrity, a tyranny of the majority in which individual liberties are denied and suppressed. On the other hand, Lefebvre recognizes that as capitalism deepens and promotes its phony spirit of individuality, as people get divided around class, and as money mediates their lives, abstract falsehoods increasingly become voluntary and instinctive, "sacred" truths nobody recognizes as myths, let alone the mythmakers.

Those who espouse a private consciousness, who flaunt it, who believe it as gospel, are, Lefebvre says, not only mystified and deluded but also *more susceptible to cults of personality, to demagogues who promise to uphold individual liberty while secretly plotting to take it away*. Lefebvre sniffs out Dostoevsky's "Grand Inquisitor" from *The Brothers Karamazov:* "Today, people are more persuaded than ever that they are completely free," says the Grand Inquisitor, "yet they have brought their freedom to us and laid it humbly at our feet."[24] People are apparently prepared to forsake their freedom in return for (national) security and happiness. They are, says the Grand Inquisitor, willing to entrust their consciences to the "captive powers" of three formidable forces: "miracle, mystery and authority."

Ironically, as Lefebvre sees it, in societies where individualism and market reification reign, an "opium of people" will flourish. A private consciousness deprived of the means to comprehend critically the broader social and political context of its own consciousness will always be manipulable and vulnerable to modern-day Grand Inquisitors. Furthermore, "religion is longer the unique 'opium' of the people," he says, "granted one tries by every means to augment the consumption of this product; there are other poisons, even more virulent; there are also circus wars and fascist buffoons. And, moreover, there are more circuses than bread. …

Modern man, for whom illusions are everywhere, can no more be simply compared to a man in a boat who believes that the horizon is moving around his vessel; he is more like a man who sets sail in a boat he believes will never be shipwrecked and it's the objects around him that toss and turn while he himself is fixed firmly on solid ground."[25]

So many illusory ideas and falsifications, *La Conscience Mystifiée* argues, so many mechanisms for upholding a *conscience privée,* are rooted in the "obscure zones" of capitalist everyday life, in actions and thoughts that become routinized and rendered "normal." "The kernel of direct, qualitative and relatively authentic human relations is," Lefebvre notes, "overwhelmed by diverse pressures. Instruments of information (TV and radio), as well as the press, consciously or not, pursue this task of investing in the sphere of deprived consciousness, exploiting it, rendering what was already deprived more deprived, bringing an illusory view of the social whole, one where deprivation has apparently disappeared. … Herein the 'socialization' of the '*conscience privée*' is pursued."[26] The fetishism of the everyday marketplace, Lefebvre warns, leads to other fetishisms, to other kinds of abstractions. Minds that are already reified are ill equipped to fend off other reifications and illusory dogmas: "The reality attributed to an abstract entity accompanies the reality attributed to the commodity."[27]

* * *

Lefebvre's Marxist voice was unusual for his generation because he cared about real individuality, about real individual freedom. He concurs with Marx's proclamation from *The Communist Manifesto* that the "free development of each is the condition for the free development of all." Consequently, it's possible to read *La Conscience Mystifiée* as much as a paean for the *individual free spirit* as an endorsement of the revolutionary collective. Lefebvre's notion of socialism plainly revolves around an association in which *dealienated* individuality can prosper within a democratic community. He isn't a socialist who makes a simple, facile dichotomy between a "good" public–collective ethos and a "bad" individual–private one. Fully developed individuality, Lefebvre argues, comes about through unfettered practice, not through drudge or routine or through uncritical enslavement to a group dogma, be it God, fatherland, or party. Capitalism has created a culture in which real liberty and community have perished behind the "free" space of the world market. And rather than drown in "the most heavenly ecstasies of religious fervor" (as Marx said in *The Communist Manifesto*), ruling classes have devised ways to mobilize heavenly ecstasies, to exploit them, to use them for their own political and economic ends.

The figure of Friedrich Nietzsche (1844–1900), who plunged into this foggy modern labyrinth, is vital here, and it is he who loiters in the foreground of *La Conscience Mystifiée* as the nemesis of mystified individuality. Lefebvre's intellectual fascination with the notorious German sage, who during the late 1930s was seen more and more as Hitler's man, took hold in the immediate aftermath of *La Conscience Mystifiée*. As fascist flames engulfed Europe, Lefebvre recalls his "necessary" rediscovery of Nietzsche, a rediscovery that culminated with what is really a continuation of *La Conscience Mystifiée,* a sort of *conscience claire,* titled simply *Nietzsche* (1939).[28] The text spans Lefebvre's early hopes with the Popular Front and culminates with the outbreak of war. Who better, he says, can help us bask in joy and burst out of misery?

Some of the most romantic pages of *La Somme et le Reste* cover Lefebvre's Nietzsche years (1936–39), years when his Nietzsche monograph unwittingly fermented. Teaching in a *collège* at Montargis, he "reread a Nietzsche never abandoned."[29] The success of the Popular Front, Lefebvre says, had been "a crowning, extraordinary success, a dazzling example of a just political idea. There was a drunkenness soon followed by a hangover."[30] Yet while he partook in the demonstrations and politicking, he distanced himself from the euphoria, from the frenzied celebrations, and found a quieter "joie de vivre" with local youth movements, frolicking, as he notes (pp. 464–65), in Fontainebleau's forest at midnight, promenading at dawn near Recloses, sleeping rough in barns in a short-lived age of innocence. "Everything became possible. All was permitted," he writes of this period (p. 464). In between, he needed his solitude, and Nietzsche became a source of comfort, something personal rather than intellectual. "Nietzsche furnished me with a system of defense. … I became a character, my character" (p. 476).

Lefebvre was seduced—and unnerved—by Nietzsche's poetry, by his ability to "think in grand, dramatic images, and his cosmic tendency."[31] How was it possible that an immense poet, "with a sonority of a grand organ, was the real ancestor and prophet of racism and Hitlerian brutality"? How could a thinker who loved the Old Testament, who went out of his way to express admiration for Jews and disdain of his German heritage, be appropriated by the National-Socialists? Why abandon Nietzsche to fascists? It was time, Lefebvre says, for the Popular Front to reclaim him. "It was also the occasion," he writes (p. 468), "to say that the political revolution, even where it might take place, wouldn't resolve all the problems of individual life, nor of love and happiness." Lefebvre's *hot* personal bond with Nietzsche also set an example for Marxists: "why," Lefebvre inquires, "should relations of Marxists to the works of Marx be so cold, so devoid of passion, so without warmth, like relations with an object?" (p. 476).

In the mid-1930s, Lefebvre propelled a personal bond into a critical political necessity, into a Nietzschean humanism, a ballast to an emergent Zeitgeist. With Nietzsche's aid, he sought to devalue bourgeois values and invent new values, stronger values, without God or nation, state, or commodity reification. At a time when reification is everywhere, when God is on the comeback and wacky fundamentalist values are writ large, a healthy dose of Lefebvre's Nietzschean skepticism strikes as piquant—and necessary. Nietzschean free spirits are fearless, strong, and secular; they don't relinquish anything to institutions or higher powers, to faith or morality. They signify the "twilight of the idols" and bid farewell "to the coldest of cold monsters."

"The coldest of cold monsters" was Nietzsche's label for the state from *Thus Spoke Zarathustra* (see "On the New Idols"). The Bush administration wasn't Nietzsche's target, but it could easily have been: "the state lies in all the tongues of good and evil; and whatever it says it lies—and whatever it has it has stolen. … Coldly, it tells lies; and this lie crawls out of its mouth: 'I, the state, am the people.' … Behold, how it lures them, the all-too-many—and how it devours them, chews them … thus roars the monster."[32] Yet Nietzschean intellectuals don't hang around to suffocate in the stench: they break windows and leap into the open air. "The earth is free even now for great souls," says Nietzsche. "A free life is still free for great souls. … Only where the state ends, there begins the human being who is not superfluous. … Where the state *ends*—look there, my brothers! Do you not see it, the rainbow and the bridges of the superman?"[33]

Being drawn to Marx and Nietzsche, Lefebvre hints in *La Somme et le Reste,* is a push-pull affair, a restless shifting between two poles, defined by will and prevailing politics. Some combination of Nietzsche and Marx can reveal a lot about the world outside and inside our heads. It's a complex connection, a veritable dialectic that dramatizes a creative tension: it seems a potentially fruitful combination unifying private consciousness and social consciousness. Rooting for Marx and Nietzsche is to support negativity, is to posit the power of the negative, to rally around it. Negating the present; overcoming the past; reaching out for the future; destroying idols, cold monsters—that's Marx's and Nietzsche's clarion call, just as it was Lefebvre's. And didn't Hegel suggest in *The Phenomenology of Spirit* that "the supreme power of being" came from "looking the negative in the face and living with it"? Power, of course, is upfront and very explicit in Nietzsche's œuvre; in Marx's it's implicit almost everywhere. His parable at the beginning of "The Working Day" chapter of *Capital* has a surprisingly Nietzschean conception of power, something not lost on Lefebvre.

"The capitalist maintains his rights as a purchaser when he tries to make the working day as long as possible," writes Marx, "and, where possible, to make two working days out of one. On the other hand, the peculiar nature of the commodity sold implies a limit to its consumption by the purchaser, and the worker maintains his right as a seller when he wishes to reduce the working day to a particular normal length."[34] Marx's verdict on the resolution strengthens his allegiance with Nietzsche: "There is here therefore an antimony of right against right and between equal rights force decides. Hence is it that in the history of capitalist production, the determination of what is a working day presents itself as a result of a struggle, a struggle between collective capital, i.e., the class of capitalists, and collective labor, i.e., the working class." Those who *win,* in other words, those who are the victors in political contestation, thereby decide what a "just" working day might be, in a principle that not only holds at the workplace but pervades every aspect of social, political, and legal life under capitalism.

In this clash of rights, Marx cautions, there's no absolute truth "out there" waiting to be discovered, no "universal" morality to be invoked to separate good from evil, right from wrong. Instead, his notion of truth mimics Nietzsche's "perspectival" notion of truth: it presents itself as a relative creation, as a clash of perspectives whose outcome emerges through struggle, through the implementation of force and power. In *The Will to Power,* Nietzsche said (section 534), "The criterion of truth resides in the enchantment of the feeling of power." And elsewhere (section 552), " 'Truth' is therefore not something there, that might be found or discovered—but something that must be created and that gives a name to a process, or rather a will to overcome that has in itself no end. … It is a word for the 'will to power.' "[35] Struggle sanctions truth claims, both Marx and Nietzsche concur: power is the judge and jury of philosophical wisdom. And everyday life is its supreme court.

There's thus no reason, Lefebvre thinks, why a Nietzschean will to power can't inspire the weak as well as the powerful, drive stiffs as well as big chiefs; a subordinate minority who's effectively a quantitative majority can strengthen their will and develop their own will to power, a will to *em*power. Here, through struggle and confrontation, through spontaneous and organized contestation, new truths about the world can be revealed and invented, those that revalue existing values and negate mystified and eternal notions. And the belief that problems of humanity are solvable through practical force, rather than abstract reasoning, seems entirely consistent with Marx's practico-critical tenets: "The question whether objective truth can be attributed to human thinking," Marx said in his second *Thesis on Feuerbach,* "isn't a question of theory but is a *practical* question. Man must prove the truth, that is, the reality and power, the one-sidedness of his thinking in practice."

Hence, in a sparkling *dénounement* to *Nietzsche,* subtitled "Nietzsche and Hitler's Fascism," Lefebvre announces a militant call to arms, releasing his own Nietzschean–Marxist will to power: "Marxists must become warriors," he urges, "without adopting the values of war."[36] In *Nietzsche,* Lefebvre smashes the windows and encourages free spirits to leap out into the fresh air. The epoch that put to bed *La Conscience Mystifiée,* yet awakens in our own, was Wagnerian not Nietzschean, Lefebvre concludes. "Nietzsche didn't love the masses. The fascists flattered the masses so much as to ensure they stayed in the situation of the masses."[37] The Nietzschean ideal of the future is in no way fascist: "His goal to overcome biological man and the man of today is an imperative precisely the contrary of a fascist postulate, after which conflicts are eternal and problems don't have a human solution. Nietzsche wouldn't have been able to support Hitler's ideology: his historical 'rumination' of the past, his cult of the state, the disdain for universalization of the individual." Consequently, "it's absurd to write Nietzsche contra Marx."[38]

In *La Somme et le Reste,* Lefebvre said nothing had happened to dampen and render unsupportable the stirring final passages of his Nietzsche book. Nothing, too, takes away from their urgency in our own decadent age, which, as Fritz Stern hinted in the *New York Times,* is drowning in "passive nihilism," a nihilism symbolizing "a decline and recession of the power of the spirit."[39] "A real culture," Lefebvre repeated in 1958 what he'd first written in 1939, "is at once a mode of living, a way of thinking and ability to act. It is a sentiment of life incorporated in a human community. It involves a relationship of human beings to the outside world. The grand culture to follow ought to integrate the cosmic into the human, instinct into consciousness. It will herald the culture of *l'homme total* [the total man], which integrates itself naturally within the Marxist conception of humanity."[40] That Lefebvre could invoke utopian man during one of the bleakest points of human history is extraordinary and inspirational for our own dark times. Nietzsche's cosmic ideal, he says, can become a socialist ideal only when it comes down to earth, where things are brutal and raw, mystified and practical. Nietzsche's *übermenschen* show real guts only when they become *menschen*—everyday people, who've descended from their Zarathustrian mountaintops, stripped away all alienations, shrugged off institutions and the state, and announced in public that God is dead—that we killed him. From then on, from an ordinary patch on planet earth, we can surge upward, breathe in the sunshine, open ourselves, come alive again. The will toward the total man marks the beginning, not the end, of history and geography.

Afterword: The End of History or the "Total Man"?

Bitterness and relief at the end. We would be able to continue for a lot longer still. We have so much more to say. We sometimes seemed to let slip something essential. Why have we consecrated the last discussion to utopia? Is it by our unanimous taste for paradox and challenge? No, it's rather because each one of us knows that all projects, theoretical or otherwise, slumber into boredom if they don't comprise a utopian dimension.

It was, in the final analysis, a marvellous autumn.

—**Henri Lefebvre,** *La révolution n'est plus ce qu'elle était*

The "total man" was nothing less than the realization of Henri Lefebvre's metaphilosophy: a Dionysian who's free and smart, versatile and sensual, who's peeled back the multiple layers of capitalist mystification and commodity reification, and who knows not only his real self but also his real relations with fellow human beings. Imagine the limit to infinity, Lefebvre urges us, a blurry figure on a distant horizon, beyond our present purview, perhaps beyond anything we've yet imagined. Here is a man and a woman separating who we are from what we might be. The total man represents a goal, an ideal, a possibility, not a historical fact; it may never become an actual fact. It comes, if it comes, without guarantees, giving "direction to our view of the future, to our activities and our consciousness."[1] It symbolizes a route open to active human practice, to thought and struggle, to striving and praxis "subjectively" overcoming "objective" conditions in the world. Nothing is assured or definitive, predestined or certain; the totality of the total man is an "open totality." The total man shouldn't be confused with the happy, smiling "new man" depicted in Socialist-Realist art, toiling for the state, somebody who's suddenly burst forth into history, complete and ready-made like a TV dinner, "in possession of all hitherto incompatible qualities of vitality and lucidity, of humble determination in labor and limitless enthusiasm in creation."[2]

The total person is "all Nature," says Lefebvre; everything lies within the grasp of this supercharacter, within this superman and superwoman who contain "all energies of matter and of life," as well as the whole past and future of the world. They're the conscience of a world gone haywire, intent on destroying itself, cannibalizing itself. Science has split the atom, propelled us to the moon, pioneered genetic engineering—and yet, we insist on truncating ourselves, impoverishing ourselves, exploiting one and another, warring and wasting vital powers, a life force hell-bent on death and annihilation. The total man approaches us from ahead, as our nemesis, looking back over his shoulder, justifiably wary and even a little incredulous. Can we raise our heads and look him in the eye? Do we have the courage to commune with him across the abyss? Lefebvre hopes we can. "Even today, at a time when our domination over Nature is already great," he writes near the end of *Dialectical Materialism* (1939), on the brink of total war, conscious that his own days may be numbered, "living man is more than ever the victim of the fetishes he himself has raised up, those strange existences, both abstract and real, brutally material yet clad in ideologies that are alluring and sometimes even bewitching. A new consciousness is needed, tenacious and skeptical, in order that these fetishes should be unmasked."

* * *

It's hard to imagine how Lefebvre would believe that the culture and society we have before us is as good as it gets. There's always something more to add, he would have insisted, always other possibilities, openings, moments of opportunity out there, on the horizon, over the rainbow. We would be able to continue for a lot longer if we could. Toward the end of his life, in 1991, as he sat in an armchair in his old house at Navarrenx, with a rug over his legs and a cat on his lap, he still wanted to talk about utopia, about the future. "We've discredited utopia," he said. "One needs to rehabilitate it. Utopia may never realize itself; and yet it is indispensable for stimulating change. Utopia is a function and a capacity, even, above all, if it doesn't realize itself. The dream of an egalitarian society, a society of abundance, is within reach though it eludes us. … But it resides there nonetheless as a means of stimulation."[3]

I remember, too, that first and only time I'd seen Lefebvre, on the TV, with Bernard-Henri Lévy, all the while telling his interlocutor he'd much rather talk about the future than the past. Perhaps he knew then; perhaps, after the Berlin Wall hadn't long tumbled down, he knew every capitalist punter would soon wallow in the glory of its demise. Perhaps Lefebvre knew, near his own end, that without some sense of utopia we'd all be lost, as a seventy-year bad rap would soon become a spectacular media bonanza. A new spirit of freedom seemed to be dawning, and now we're living in its scary midst. Perhaps he'd suspected as much. I'd little realized back then—couldn't realize—how Lévy's program *The Spirit of Freedom* and the companion book *Les Aventures de la Liberté* set the tone for the shallowness and narrowness the new century would come to epitomize. Punctuated by subheadings like "The Great Hopes," "Times of Contempt," "Lost Illusions," and "The End of the Prophets," the text's cynicism reeked: give up the ghost, abandon all hope ye who enters here.

Around the same time as *Les Aventures de la Liberté* hit French bookstores and around the time *Le Monde* announced Lefebvre's death—the death of a *style*—across the Atlantic another scurrilous book by Francis Fukuyama danced to a similar refrain: "the end of history."[4] Extending an article-length thesis that had aired a few years earlier in the conservative *National Interest,* Fukuyama flagged up "the end point of mankind's ideological evolution … the final form of human government": liberal bourgeois democracy. We've reached the moment, Fukuyama bragged, of "remarkable consensus." Liberal democracy had won its legitimacy, conquering all rival ideologies, and, he thought, we should be glad. Hereditary monarchy had run its course a while back, and so had fascism; and now, apparently, so had communism. There's no other tale to tell, no alternative, no other big idea left, nothing aside from bourgeois democracy and free-market economics. It was totalitarian now even to *think* about other big ideas about human progress. The year 1991 heralded, in the infamous words of George Bush, Sr., "a New World Order."[5] "We cannot picture to ourselves," Fukuyama proclaimed in *The End of History and the Last Man,* "a world that is *essentially* different from the present one, and at the same time better. Other, less reflective ages also thought of themselves as the best, but we arrive at this conclusion exhausted, as it were, from the pursuit of alternatives we felt *had* to be better than liberal democracy."

Yet the Stalinist One-State we once knew over there has since come home to roost here, in the West, in the guise of a new Washington consensus that lies, cheats, and bullies its way to capitalist fame and glory. Never has mediocrity reached such dizzy heights of power and wealth; never has deceit and corruption been part of its political arsenal. The dogmatism Lévy and Fukuyama tag on the twentieth-century tradition of socialism pales alongside the false testimonies and propaganda pervading every aspect of daily life today. Beset by conflict, crisis, war, terrorist threat, and fundamentalism of every stripe, the legitimacy of liberal democracy has never looked so extraordinarily fragile. The tragedy is palpable. Truth and falsity have degenerated into interchangeable language games, fair game for the rich and powerful, for those who control the media. Fukuyama's belief that liberal democracies have less incentive for war, and have universally satisfied people's need for reciprocal recognition, seems even more ridiculous than it did a decade ago.

More recently, Fukuyama has been struggling for his own recognition against a neoconservative backlash, with a few utopian ideas of its own.[6] The ideological prophet of Poppy Bush's "New World Order," an order that heralded the "last man," the happy (mystified?) citizen whose "long-run" interests were apparently fulfilled, now distances himself from the reality of a state he'd once affirmed as incarnating universal liberty. Perhaps history has opened up again? Or maybe George W. is just a historical blip? But Fukuyama can't have it both ways in his Bush critique: "In order to refute my hypothesis," he wrote in his original *National Interest* article, with a typical spirit of mild-mannered closure, "it is not sufficient to suggest that the future holds in store large and momentous events. One would have to show that these events were driven by a systematic idea of political and social justice that claimed to supersede liberalism." Charles Krauthammer, a conservative columnist for the *Washington Post,* plainly believes the future is still there for the taking; the complacent "The End of History" honeymoon is over.

"Democratic realism," says Krauthammer, is what American foreign policy calls for: the Right should reclaim the utopian spirit for itself and make it real, project through military might its conservative values across global space. If the masses can be kept mystified at home, neocon power elites can produce space abroad—and control the world. A new inner and outer dialectic infuses Lefebvre's theory of capitalist domination and expansion. "The 1990s were a holiday from history," Krauthammer writes,

> an illusory period during which we imagined that the existential struggles of the past six decades against various totalitarianisms had ended for good. September 11 reminded us rudely that history had not ended, and we found ourselves in a new existential struggle, this time with an enemy even more fanatical, fatalistic and indeed undeterable than in the past. Nonetheless, we had one factor in our favor. With the passing of the Soviet Union, we had entered a unique period in human history, a unipolar era in which America *enjoys* a predominance of power greater than any that has existed in the half-millennium of the modern state system.[7]

The offensive edge to the cybernanthrope's world order, bolstered by high-tech weapons of mass destruction and distraction, schemed in think tanks like the American Enterprise Institute, is a grave threat to our collective future. What Lefebvre bequeaths us is a theoretical apparatus that helps us demystify these machinations, probe into this dark Dr. Strangelove labyrinth, explain its logic, and understand its mentality, in all its madness. Meanwhile, his legacy equips us with a youthful spirit of *confrontation:* a battle around not only ideas and scholarly critique but also political confrontation, spontaneous and practical confrontation out on the streets, making noise and demanding one's rights, fighting the power, and looking the negative in the face and struggling against it. Lefebvre's whole life and œuvre evolved and flourished through confrontation: confrontation with the Surrealists, confrontation with the Situationists, confrontation with the Communist Party, confrontation with his Catholic faith, confrontation with his Pyrenean roots, confrontation with fascism, confrontation with Hitler, confrontation with the past as well the future, confrontation with himself and his world.

Lefebvre lived though a century of madmen and dictators, defeats and disasters, crises and conspiracies, and he can help us confront the demons that haunt our new century, ones that "*enjoy* the predominance of power." "Men can and must set themselves a total solution," he insisted in 1939, just when everything seemed lost. "We don't exist in advance, metaphysically. The game has not already been won; we may lose everything. The transcending is never inevitable. But it is for this precise reason that the question of man and the mind acquires an infinite tragic significance, and that those who can sense this will give up their solitude in order to enter into an authentic spiritual community."[8]

* * *

One of the last books Lefebvre read—reread—was by Franz Kafka: *The Castle*.[9] It was a book he thought particularly pertinent for the present conjuncture, where castles and ramparts reign over us all, in plain view, but cut off somehow, and occupants are evermore difficult to pin down when we come knocking on their doors, providing we can find the right door to knock on. Following Kafka, Lefebvre perhaps recognized the thoroughly modern conflict now besieging us, a conflict not of us against other men but of us against a world transformed into an immense administration.

The shift Kafka made between his two great novels, *The Trial* (1917) and *The Castle* (1922), makes for a suggestive shift in our administered world. In *The Trial,* Joseph K., like a dog, stands accused in a world that's an omnipotent tribunal, a sort of state-monopoly capitalist system. In *The Castle,* the protagonist K. populates a world that's shrunken into a village whose dominating castle on the hill seems even more powerful and elusive than ever. Perhaps Henri—or H., as we could call him—saw that village as our new "global village," a world shrunken by globalization, wherein a psychological drama of one man confronting this castle is really a political parable of us, today, having to conceive a collective identity—to confront the abstract, gothic mystery we ourselves have created. "Direct intercourse with the authorities was not particularly difficult," K. muses,

> for well organized as they might be, all they did was guard the distant and invisible interests of distant and invisible masters, while K. fought for something vitally near to him, for himself, and moreover, at least at the very beginning, on his own initiative, for he was the attacker. … But now by the fact that they had at once amply met his wishes in all unimportant matters—and hitherto only unimportant matters had come up—they had robbed him of the possibility of light and easy victories, and with that of the satisfaction which must accompany them and the well-grounded confidence for further and greater struggles which must result from them. Instead, they let K. go anywhere he liked—of course only within the village—and thus pampered and enervated him, ruled out all possibility of conflict, and transported him into an unofficial, totally unrecognized, troubled, and alien existence. … So it came about that while a light and frivolous bearing, a certain deliberate carelessness was sufficient when one came in direct contact with the authorities, one needed in everything else the greatest caution, and had to look round on every side before one made a single step.[10]

Almost a century on, progressives need the greatest caution in everything we do; we need to look around on every side before we can make a single step. The gravity of the situation isn't lost on any of us. And yet, at the same time, there's a sense that we should, and can, lighten up. After all, even amid the existential no-exits of Kafka, a black humor radiates, a glint of light warms a cold corner: as Kafka's fellow countryman Milan Kundera notes in his latest book *Le Rideau* [The Curtain], Kafka "wanted to descend into the dark depths of a joke [*blague*]."[11] It was comedy that let K. deal with tragedy and let him pull back the curtain, rip it down, and tear it apart. He can still help us see what lies inside and beyond the wrapping, and H. knew it. Indeed, Kundera's metaphor seems apt for H., who ripped down curtains suspended in front of our Kafkaseque modern world, *de*masked them, named what lay behind them, and asked us to look within.

Lefebvre's most Kafkaesque book is *Vers le cybernanthrope* (1971), where H. became a land surveyor facing the cybernanthrope's tribunal, trapped within the confines of his rational castle, searching for a way out, confronting curtains of systematized mystification. In its corridors, the cybernanthropic last man stalks the Lefebvrian total man in a duel over our collective destiny. But it's humor that will win out in the end. The cybernanthrope, H. says, is neither tragic nor comical: he's farcical. He's a product of a farcical situation and farcical events. Of course, he doesn't see himself as farcical, because he's rather earnest, taking seriously his duties, his realism. What's in store for us, H. thinks, is another world war, a guerilla war that any potential total man needs to keep on waging, using as arms spirit and satire. We'll have to be perpetual inventors, H. says, restless creators and re-creators. We'll have to cover our tracks, engage in pranks and jokes, knock cybernanthropes off balance, keep them guessing. For vanquishing, for even engaging in battle, we'll valorize imperfections and disequilibria, troubles and gaps, excesses and faults. We'll valorize desire and passion, and revel in irony and comedy. We'll use slingshots against tanks, nets against armor, clatter against chatter. We will, H. assures us, vanquish by *style,* a style of grand negativity and absolute subversion, of critical engagement and mocking revolt.

It's a style that can never entirely go out of fashion. Not quite.

Notes

Preface

1. "BHL," his familiar media acronym, is famous as much for his flowing locks, on-screen arrogance, and unbuttoned shirts as he is for the rigor of his thought. As Perry Anderson recently lamented, "The general condition of intellectual life [in France] is suggested by the bizarre prominence of Bernard-Henri Lévy, far the best-known 'thinker' under 60 in the country. It would be difficult to imagine a more extraordinary reversal of national standards of taste and intelligence than the attention accorded this crass bobby in France's public sphere, despite the innumerable demonstrations of his inability to get a fact or an idea straight." (See Perry Anderson, "Dégringolade," *London Review of Books,* September 2, 2004.)
2. The whole series has been transcribed and translated into English. See Bernard-Henri Lévy, *Adventures on the Freedom Road* (Harvill, London, 1995). The vignette on Henri Lefebvre appears in the chapter titled "A Group of Young Philosophers: Conversations with Henri Lefebvre." This citation is found on p. 131. See also Bernard-Henri Lévy, *Les Aventures de la Liberté* (Grasset, Paris, 1991).
3. Henri Lefebvre, *La Somme et le Reste—Tome II* (La Nef de Paris, Paris, 1959), p. 512.
4. Olivier Corpet, "Les Aventures d'un Dialecticien," *Le Monde,* July 2, 1991.
5. Eric Hobsbawm, *Age of Extremes: The Short Twentieth-Century, 1914–1991* (Abacus, London, 1995). A brilliantly erudite synthesis of the anecdotal and the structural, a chip off Lefebvre's own block, Hobsbawm's personal vision of the past century actually identifies 1991 as its cutoff point.

6. Manuel Castells, "Citizen Movements, Information and Analysis: An Interview with Manuel Castells, *City* 7 (1997): 146.
7. *La Somme et le Reste—Tome I* (La Nef de Paris, Paris, 1959), p. 46.
8. Jean-Paul Sartre, *Search for a Method* (Vintage, New York, 1968), p. xxxiii. Sartre admired Lefebvre's dialectical method: "Yet it is a Marxist, Henri Lefebvre," Sartre wrote in *Search for a Method* (pp. 51–52), "who in my opinion has provided a simple and faultless method for integrating sociology and history in the perspective of a materialist dialectic. … We only regret that Lefebvre has not found imitators among the rest of Marxist intellectuals." The compliment, though, wasn't reciprocated. Always incredulous of Sartrean existentialism, especially in its implications for individual freedom and political action, Lefebvre penned a fierce diatribe contra Sartre in *L'existentialisme* (Éditions du Sagittaire, Paris, 1946). Even Lefebvre's staunchest fans agree this text is best forgotten.
9. *La Somme et le Reste—Tome II* (Bélibaste Éditeur, Genève, 1973), p. 11. In the same text (chapter III), Lefebvre actually defends Proust against his socialist detractors. "One can only wish," Lefebvre wrote (p. 41), "that some day socialist realism can attain an art as subtle as Proust's, and reach a similar visionary power and Romanesque construction."
10. *Conversation avec Henri Lefebvre* (Messidor, Paris, 1991), p. 22.
11. Lefebvre, *La Somme et le Reste—Tome II,* p. 676.
12. Cited in Edward Soja, *Thirdspace* (Blackwell, Oxford, 1996), p. 33.
13. *Critique of Everyday Life—Volume 1* (Verso, London, 1991), p. 202.
14. See *Conversation avec Henri Lefebvre,* p. 98.
15. Ibid., p. 99.
16. Ibid., p. 99.
17. *La Somme et le Reste—Tome I,* p. 242.
18. Ibid.
19. *Conservation avec Henri Lefebvre,* pp. 97–98.
20. Henri Lefebvre, *Pyrénées* (Éditions Rencontre, Lausanne, 1965), p. 10.
21. Ibid., p. 10.

Chapter 1

1. Immediately after the war, Lefebvre taught philosophy at a *lycée* (where students prepare for their preuniversity *baccalauréat*) in Toulouse and at a local military college. (One can chuckle at the young cadets being drilled with dialectical materialism.) At the same time, with a little help from his old Dada friend Tristan Tzara, whom he'd known since his mid-1920s university years at the Sorbonne, Lefebvre broadcasted on Radio Toulouse. No source indicates what he spoke about there or whether any tape recordings of the broadcasts survive him. It's likely he spread Communist Party wisdom in the postwar reconstruction period, when the political stakes were high. Afterward, long-time Lefebvre friend and supporter Georges Gurvitch, the sociologist and Russian revolutionary exile, engineered Lefebvre's tenure as a researcher at CNRS in Paris from November 1947.

2. *La Somme et le Reste—Tome II,* p. 464.
3. See Remy Hess, *Henri Lefebvre et l'aventure du siècle,* p. 110; cf. *Conversations avec Henri Lefebvre,* pp. 50–51.
4. *Conversations avec Henri Lefebvre,* p. 50.
5. Ibid., pp. 50–51.
6. Victor Serge, *Memoirs of a Revolutionary, 1901–1941* (Oxford University Press, Oxford, 1963), p. 362.
7. As Michel Trebitsch pointed out in his preface to *Critique of Everyday Life—Volume 2* (Verso, London, 2002), p. xiv, this was Lefebvre's first academic qualification since his *diplôme d'études supérieures,* obtained in 1924. It gives hope to every aging grad student! Lefebvre's secondary thesis, a necessary component of the French educational system, was called *La Vallée de Campan;* Presses Universitaires de France published it in book form in 1958. The historian Michel Trebitsch prefaced all three volumes of Lefebvre's *Critique of Everyday Life;* sadly, in March 2004, cancer saw off at age fifty-five one of the most gifted francophone interpreters and disseminators of Lefebvre.
8. Jean-Paul Sartre, *Search for a Method,* p. 52 (emphasis in original).
9. *Critique of Everyday Life—Volume 1,* p. 129.
10. Ibid., p. 216.
11. Between 1947 and 1955, Lefebvre absorbed himself in French literature, consecrating a series of books on classical masters like Denis Diderot (1949), Blaise Pascal (two volumes, 1949 and 1955), Alfred de Musset (1955), and François Rabelais (1955). He used these writers not only to sharpen the concept of everyday life but also as antidotes to bourgeois values and party dogma. Lefebvre's "retreat" into literature during this period seemed positively correlated with his growing alienation from the party.
12. *Critique of Everyday Life,* p. 14.
13. Ibid., p. 11.
14. Ibid., p. 27. Lefebvre devotes more attention to Joyce's *Ulysses,* and to *Finnegans Wake,* early on in *Everyday Life in the Modern World* (Penguin, London, 1971); see pp. 2–11.
15. James Joyce, *Ulysses* (Penguin, Harmondsworth, 1986), pp. 643–44.
16. The dramas around daily bread and French village life, together with all their comical hypocrisies and shenanigans, are mischievously evoked in Marcel Pagnol's classic 1938 film, *La Femme du Boulanger* [The Baker's Wife]. See the film's text, whose dialogue is in the broken language of real, everyday people (*La Femme du Boulanger* [Presses Pocket, Paris, 1976]).
17. Pierre Mac Orlan, *Villes* (Gallimard, 1929), p. 247.
18. *Critique of Everyday Life,* p. 48.
19. The desire also conjures up the spirit of the late Isaac Babel, the Ukrainian short-story wizard carted off by Stalin's henchmen one dark night in May 1939, never to return. "The Party, the government, have given us everything," Babel said, "depriving us only of one privilege—that of writing badly!" See Isaac Babel, *The Lonely Years, 1925–1939* (Farrar, Straus and Co., New York, 1964), p. 399.

20. *Critique of Everyday Life,* p. 49.
21. Ibid., p. 13.
22. Lefebvre, *Everyday Life in the Modern World* (Penguin, London, 1971), p. 14.
23. *Critique of Everyday Life,* p. 6.
24. Ibid., pp. 37–38.
25. Karl Marx, "The Economic and Philosophical Manuscripts," in *Karl Marx—Early Writings* (Penguin, Harmondsworth, 1974), p. 326.
26. Marx, *Capital—Volume 1* (Penguin, Harmondsworth, 1976), p. 493.
27. Karl Marx and Frederick Engels, *The Communist Manifesto* (Verso, London, 1998), p. 38.
28. *Critique of Everyday Life,* p. 202.
29. Ibid., p. 207. Emphasis in original.
30. The urging is Rabelais's opening address "To My Readers." I've cited Burton Raffel's translation of *Gargantua and Pantagruel* (W.W. Norton, New York, 1990).
31. Lefebvre, *Rabelais* (Anthropos, Paris, 2001), p. 213.
32. Mikhail Bakhtin, *Rabelais and His World* (Indiana University Press, Bloomington, 1984), p. 10.
33. Rabelais, *Gargantua and Pantagruel,* p. 124.
34. Lefebvre, *Rabelais,* p. 112.
35. Rabelais, *Gargantua and Pantagruel,* p. 124.
36. Lefebvre, *Rabelais,* pp. 113–14.
37. Ibid., p. 203.

Chapter 2

1. In the mid-1980s, just before soaring rents forced Lefebvre out of rue Rambuteau, he wrote a quirky "Rhythmanalysis" essay titled "Seen from the Window," describing the rhythms, murmurs, and noises of the street down below. "From the window opening onto rue R.," he says, "facing the famous P. Centre, there is no need to lean much to see into the distance. … To the right, the palace-centre P., the Forum, up as far as the Bank of France. To the left up as far as the Archives, perpendicular to this direction, the *Hôtel de Ville* and, on the other side, the *Arts et Metiers.* The whole of Paris, ancient and modern, traditional and creative, active and idle" is there. See Henri Lefebvre, *Rhythnanalysis* (Continuum Books, London, 2004), p. 28. As of December 2004, the Forum, a subterranean shopping arcade once described by historian Louis Chevalier as "a deep, fetid underground," will soon be history. Paris's socialist mayor Bertrand Delanoë chose David Mangin's ecological sensitive two-hectare garden cum public square, with a giant luminous roof, as the Forum's more worthy replacement.
2. Lefebvre, *Everyday Life in the Modern World,* p. 58.
3. Ibid., p. 58.

4. "An Interview with Henri Lefebvre," *Environment and Planning D: Society and Space* no. 5 (1987): 27–38.
5. Lefebvre, *Critique of Everyday Life—Volume 2* (Verso, London, 2002), p. 3.
6. Ibid., p. 41.
7. Ibid., p. 89.
8. See David Riesman, Nathan Glazer, and Reuel Denney, *The Lonely Crowd: A Study of the Changing American Character* (Doubleday Books, New York, 1953).
9. William H. Whyte, *The Organization Man* (Simon and Schuster, New York, 1956).
10. Herbert Marcuse, *One Dimensional Man: The Ideology of Industrial Society* (Routledge and Kegan Paul, London, 1964). Alongside Max Horkheimer and Theodor Adorno, Marcuse was one of the pioneers of the celebrated "Frankfurt School of Social Research," a left-wing think tank that worked on critical theory, aesthetics, and politics. In attempting to figure out modern industrial and technological society, new state forms, and ideological manipulation, they brought Marx, Hegel, and Freud into an imaginative dialogue. In the 1930s, the School, dominated by Jews, was forced to quit Germany; Marcuse bivouacked for years in the United States and taught philosophy at the University of California, San Diego.
11. *Conversation avec Henri Lefebvre,* p. 70.
12. Lefebvre, *The Explosion: Marxism and the French Upheaval* (Monthly Review Press, New York, 1969), p. 31; emphasis in original. Marshall Berman, at the sprite age of twenty-three, voiced a similar critique of Marcuse's "closed, fatalistic perspective" in 1964. "Marcuse has become more concrete with advancing age, more involved than ever in the socio-pathology of everyday life. … [He] tries to explain advanced industrial society as a smoothly functioning system in which every aspect of life reinforces the others, an infernal machine in which all parts mesh to grind the spirit down. … He is not accustomed to [society's] dark and twisted ways"; Marshall Berman, "Theory and Practice," *Partisan Review* (Fall 1964): 619.
13. The butterfly figures as a powerful romantic metaphor in *La Somme et le Reste* (see, especially, *Tome II,* p. 428), even as a symbol of Lefebvre's anarchist tendencies. One incident in particular is recalled, from Lefebvre's military service in 1926. Out on an infantry exercise one early summer morning, "I glimpsed ten steps ahead of me, at the side of the lane, a lovely butterfly whose rose wings where damp; this prevented him from flying. I hastened myself, took him as delicately as possible and placed him down on the embankment." Three seconds later, a corporal sticks a rifle butt in Lefebvre's back. The captain on horseback shouts, "*Chasseur* Lefebvre! 8 days in police detention." "This lad announces himself as a dangerous subversive element … a soft dreamer, a saviour of butterflies … an intellectual anarchist."
14. *Critique of Everyday Life—Volume II,* p. 348 (emphasis in original).

15. Stéphane Mallarmé, preface to "A Roll of the Dice Will Never Abolish Chance" (1895), in *Stéphane Mallarmé—Selected Poetry and Prose,* ed. Mary Ann Caws (New Directions Books, New York, 1982), p. 105. Mallarmé's poem sprawls diagonally across the page, with certain verses interspersed with others; odd words dwell alone just as others interlock and interweave. Sometimes, you don't know whether the verses flow over the page or down the page or in both directions simultaneously.
16. Henri Bergson, *Creative Evolution* (Modern Library, New York, 1944), p. 337.
17. Ibid., p. 393.
18. *La Somme et le Reste—Tome I,* p. 234.
19. Ibid., p. 235.
20. *Critique of Everyday Life—Volume 2,* p. 347.
21. *La Somme et le Reste—Tome II,* p. 647.
22. *Critique of Everyday Life—Volume 2,* p. 345.
23. Ibid., p. 351.
24. For more details on Debord's (1931–94) stormy life and complex thought, see my *Guy Debord* (Reaktion Books, London, 2005). On the brink of insurgency, Debord published *The Society of the Spectacle* (1967), his best-known text, a work that would become *the* radical book of the decade, perhaps even the most *radical* radical book ever written. Utterly original in composition, its 221 strange, pointed aphorisms blend a youthful Marx with a left-wing Hegel, a bellicose Machiavelli with a utopian Karl Korsch, a militaristic Clausewitz with a romantic Georg Lukács. Debord reinvented Marxian political economy as elegant prose poetry, and with its stirring refrains, *The Society of the Spectacle* indicted an emergent world order in which unity really spelled division, essence appearance, truth falsity.
25. "Lefebvre on the Situationist International," *October* (Winter 1997): 69–70.
26. Lefebvre, *Le Temps des Méprises* [Times of Contempt] (Éditions Stock, Paris, 1975), p. 158.
27. Ibid., p. 151.
28. "Lefebvre on the Situationist International," p. 70.
29. Cited in Christophe Bourseiller's *Vie et mort de Guy Debord* (Plon, Paris, 1999), pp. 258–59. Nicole gave birth to Lefebvre's sixth child, daughter Armelle, in 1964. In 1978, at the age of seventy-seven, Lefebvre married Catherine Regulier, then a twenty-one-year-old communist militant. Estranged from her parents because of her relationship with Lefebvre, Catherine and Henri stayed together until the end of his life.
30. Guy Debord, "In Girum Imus Nocte et Consumimur Igni," in *Guy Debord—Œuvres Cinématographiques Complètes, 1952–1978* (Gallimard, Paris, 1978), p. 253. Debord's threnody to Paris, and his denunciation of the established film world, has a Latin palindrome title with an English translation: "We go round and around and are consumed by fire."
31. "Lefebvre and the Situationist International."
32. See Andy Merrifield, *Guy Debord,* especially chap. 1.

33. Debord, "Réponse à une Enquête du Groupe Surréaliste Belge," in *Guy Debord Présente "Potlatch" (1954–1957)* (Gallimard, Paris, 1996), p. 42.
34. Debord, *Guy Debord Correspondance, Volume 1: juin 1957–août 1960* (Librairie Arthème Fayard, Paris, 1999), p. 313.
35. Ibid., p. 318.
36. Ibid., p. 318. Emphasis in original.
37. Stendhal, *Racine and Shakespeare,* cited in Henri Lefebvre, *Introduction to Modernity* (Verso, London, 1995), p. 239. Stendhal (1783–1842) was the penname of Henri Beyle, whose romantic novels, especially *Scarlet and Black* (1830) and *The Charterhouse of Parma* (1839), brought him fame and a following. Stendhal dedicated his works to "the happy few" and coined the term *Beylism* as his philosophical credo for the pursuit of happiness. His dedication may have been an allusion to Shakespeare's *Henry V:* "We few, we happy few, we band of brothers." Interestingly, and unbeknownst to the Lefebvre of *Introduction to Modernity,* Shakespeare's phrase would feature in Guy Debord's film version of *The Society of the Spectacle* (1973). Following the caption of "we happy few," the frame flashes to wall graffiti at an occupied Sorbonne, circa late 1960s: "Run quickly, comrade, the old world is behind you!"
38. Lefebvre, *Introduction to Modernity,* p. 239.
39. Ibid., p. 346. Emphasis in original.
40. Ibid., p. 359. For his own part, Debord responded graciously to Lefebvre in a letter dated May 5, 1960. "I am counting on the perspectives of the Situationists," the Situ leader told Lefebvre, "(which, as you know, don't fear going far out) for at least reconciling romanticism with our revolutionary side; and better, for eventually overcoming all romanticism." See *Guy Debord Correspondance, Volume 1,* p. 332.
41. *Introduction to Modernity,* p. 258.
42. Ibid., p. 302.

Chapter 3

1. Interview with author, March 15, 2005. Today, Daniel Cohn-Bendit, the ex-'68 student leader, is copresident of Greens/Free European Alliance in the European Parliament. He's also a frequent (and outspoken) political commentator on French TV with left-democratic, pro-European integrationist ideals.
2. *Introduction to Modernity,* p. 343.
3. Lefebvre, *The Explosion: Marxism and the French Upheaval* (Monthly Review Press, New York, 1969), p. 7. All page citations to follow refer to this edition.
4. Lefebvre himself began to document changes (and contradictions) between "politics" and "the economy" from the mid-1970s onward in a series of volumes on the state. The title alone of one of them captured the nub of the shift away from a managerialist style of national government to a

entrepreneurial, often supranational, one: *Le mode de production étatique* (1977)—the statist mode of production.

5. Vladimir Lenin, *One Step Forward, Two Steps Back (The Crisis in Our Party)* (Progress Publishers, Moscow, 1978).
6. Rosa Luxemburg, *The Russian Revolution, and Lenin or Marxism?* (University of Michigan Press, Ann Arbor, 1961).
7. One of the great countercultural texts of Lefebvre's generation, urging the same exuberance to Rabelaisian audiences across the ocean, was Norman O. Brown's *Life against Death* (Wesleyan University Press, Middletown, Connecticut, 1959): "It was Blake who said that the road to excess leads to the palace of wisdom; Hegel was able to see the dialectic of reality as 'the bacchanalian revel, in which no member is not drunk.' … The only alternative to the witches' brew is psychoanalytical consciousness, which is not the Apollonian scholasticism, but consciousness embracing and affirming instinctual reality—Dionysian consciousness" (p. 176).
8. *The Survival of Capitalism,* p. 100. "There must be an objective," Lefebvre says, "a strategy: nothing can replace political thought, or a cultivated spontaneity." Curiously, when Lefebvre published *La survie du capitalisme* in 1973, he included several essays that had already figured in *The Explosion* [*L'irruption de Nanterre au sommet*], including "Contestation, Spontaneity, Violence." Alas, the English version removed these repetitions, denying Anglophone scholars the chance to muse on why the doubling up. The subtitle of *Survival* offers clues: "reproduction of relations of production." Five years on from '68, the capitalist system had not only withstood "subjective" bombardment but also "objectively" began to grow. The essential condition of this growth is that relations of production can be *reproduced.* How are they reproduced? In a wink to Althusser, Lefebvre's text is less exuberant in its revolutionary hopes and enters into the world of institutional analyses; yet it's obvious he can't quite resist toying with the idea of spontaneity and contestation throwing a spanner in the apparatus of societal reproduction. See, for more details, Remi Hess's enlightening "Postface" to the third edition of *La survie du capitalisme* (Anthropos, Paris, 2002), pp. 197–214.
9. See Anthony Giddens, *The Third Way: The Renewal of Social Democracy* (Polity, London, 1998).

Chapter 4

1. In *Pyrénées*, Lefebvre calls Mourenx a "semi-colony," built between 1957 and 1960 for gas workers at the plant in nearby Lacq. Of Lacq, Lefebvre notes (p. 116), "The 'complex,' according to the pompous and imprecise vocabulary of the technocrats, encrusts itself in the landscape like a foreign body." "Who had profited?" from this alien intrusion. "Before all Paris, before all private enterprise, who receive from here energy and natural resources, and who've participated in the trappings of mobilizing the gigantic means of state capitalism" (p. 117).

2. *Introduction to Modernity,* p. 118. All parenthetical page numbers henceforth refer to this text.
3. In a 1983 interview, Lefebvre noted how the Situationists devised *dérive*—collective, often nocturnal pedestrian drifts through urban space. The practice sought to expose the idiocy of an urbanism based on monofunctional separation. These drifts tapped the "psychogeography" of different neighborhoods (in Paris, London, Amsterdam) and cognitively stitched together the urban fabric by emphasizing what was getting torn apart and plundered. "We had a vision of a city," Lefebvre said, "that was more and more fragmented without its organic unity being completely shattered." See "Lefebvre and the Situationists International," *October* (Winter 1997).
4. *Introduction to Modernity,* p. 122.
5. "The abstraction of the *state as such,*" Marx wrote in his *Critique of Hegel's Doctrine of the State* (1843), "wasn't born until the modern world because the abstraction of the *political state* is a modern product" (see *Karl Marx—Early Writings,* p. 90; emphases in original).
6. For more details on the Marxist position vis-à-vis the homegrown artisan-anarchism of Pierre-Joseph Proudhon (1809–1865) see my *Metromarxism* (Routledge, New York, 2002), pp. 42–45. For Lefebvre's own views on the matter, see his stimulating discussion on Frederick Engels's *Housing Question* "Engels et l'utopie," reprinted in *Espace et Politique*. Lefebvre's latter booklet is included in Anthropos's edition of *Le droit à la ville* (Paris, 1972).
7. Cf. Henri Lefebvre, *The Urban Revolution* (Minnesota University Press, Minneapolis, 2003), p. 84; see, too, *La révolution urbaine* (Gallimard, Paris, 1970), p. 114.
8. Karl Marx, *Capital—Volume 1* (Penguin, Harmondsworth, 1976), p. 284.
9. Lefebvre's *Le droit à la ville* has been translated by Eleonore Kofman and Elizabeth Lebas and introduced in their edited *Writings on Cities—Henri Lefebvre* (Blackwell, Oxford, 1996). On occasion, I've tweaked their translation. Page references to follow use the English as well as Anthropos's original 1968 French version, *Le droit à la ville suivi de Espace et politique*. Lefebvre's urban impulse had already been glimpsed in a detailed historical account of the 1871 Paris Commune, *La Proclamation de la Commune* (Gallimard, Paris, 1965), but it was around, and especially after, 1968 that his urban œuvre really took off. Along with *Le droit à la ville,* this would include *Du rural à l'urbain* (Anthropos, Paris, 1970), *La révolution urbaine* (Gallimard, Paris, 1970), and *La pensée marxiste et la ville* (Casterman, Paris, 1972). Thus, by 1972, in his seventy-first year, Lefebvre could justifiably be called an urban scholar. His critique of "urbanism," and his analyses of "urban space," would soon edge him toward studying the role of geography in the "survival of capitalism," culminating with *La production de l'espace* (Anthropos, Paris, 1974).
10. "What relation is there today," Lefebvre asks in *The Right to the City* (p. 92), "between philosophy and the city?" An ambiguous one, he responds, unambiguously. "The most eminent contemporary philosophers," says he, "don't

borrow their themes from the city. [Gaston] Bachelard has left admirable pages consecrated to the house [in *The Poetics of Space*]. Heidegger has meditated on the Greek city and Logos, on the Greek temple. Yet the metaphors epitomizing Heideggerian thought don't come from the city but from a native and earlier life: the 'shepherds of being,' the 'forest paths.' … As for so-called 'existentialist' thought, it is based on individual consciousness, on the subject and the ordeals of subjectivity, rather than on a practical, historical and social reality."

11. Lefebvre, "Engels et l'utopie," p. 217; cf. *Metromarxism,* pp. 42–48.
12. In *The Right to the City: Social Justice and the Fight for Public Space* (Guilford, New York, 2003), the urban geographer Don Mitchell puts a provocative twist on Lefebvre's thesis, immersing it in the North American legal system. In staking out a beachhead for disenfranchised homeless people, and other expulsees from the city's public realm, Mitchell pushes a Lefebvrian right into a twenty-first-century urban ethic.
13. Cited in Remi Hess, *Henri Lefebvre et l'aventure du siècle,* p. 315.
14. Lefebvre, "No Salvation away from the Center?" in *Writing on Cities — Henri Lefebvre* trans. and repr. E. Kofman and E. Lebas, p. 208.
15. Personal communication, August 30, 2004. For other Soja insights and reminiscences of "the dear old man," see his *Thirdspace* (Blackwell, Oxford, 1996). Jameson's essay, "Postmodernism, or the Cultural Logic of Late Capitalism," *New Left Review* 146 (1984): 53–92, replete with Lefebvre's looming presence, is now a classic. A book-length version, sporting the same title, was published by Verso in 1991. "The notion of a predominance of space in the postcontemporary era we owe to Henri Lefebvre," wrote Jameson (p. 364), "(to whom, however, the concept of a postmodern period or stage is alien)."
16. See Lefebvre, *The Production of Space* (Blackwell, Oxford, 1991), pp. 73–74.
17. Ibid., p. 77.
18. "Rhythmanalysis of Mediterranean Cities," in *Writings on Cities—Henri Lefebvre,* p. 236.
19. "No Salvation away from the Center?" *Writings on Cities—Henri Lefebvre,* p. 208. Interestingly, Lefebvre's love affair with Florence and Venice was shared by Guy Debord, who made clandestine visits to Venice (including a poignant one just before his suicide in 1994) and, during the 1970s, went into "exile" for several years in Florence's Oltarno district: "There was this little Florentine who was so graceful. In the evenings she would cross the river to come to San Frediano. I fell in love very unexpectedly, perhaps because of her beautiful, bitter smile. I told her, in brief: 'Do not stay silent, for I come before you as a stranger and a traveller. Grant me some refreshment before I go away and am here no more.' " Guy Debord, *Panégyrique* (Verso, New York, 1991), p. 47.
20. *Writing on Cities—Henri Lefebvre,* p. 208.
21. Lefebvre, *Rhythmanalysis,* trans. Stuart Elden and Gerald Moore (Continuum Books, London, 2004).

22. Lefebvre, *The Urban Revolution*. Here, and in the chapter to follow, I've nudged Robert Bononno's English translation subject to my own reading of the 1970 original *La révolution urbaine*.
23. "Executive Summary," Second Session of the World Urban Forum, Barcelona, Spain, September 13–17, 2004. See www.unhabitat.org/wuf/2004/default.asp

Chapter 5

1. Lefebvre, *La révolution urbaine* (Gallimard, Paris, 1970), p. 13; *The Urban Revolution,* trans. Robert Bononno (Minnesota University Press, Minneapolis, 2003), pp. 5–6. In what follows, I cite parenthetically, using this ordering, page numbers from both editions.
2. David Harvey, *Social Justice and the City* (Edward Arnold, London, 1973), pp. 302–303.
3. Ibid., p. 303.
4. See David Harvey, *The Urban Experience* (Basil Blackwell, Oxford, 1989), pp. 59-89.
5. David Harvey, *The Urbanization of Capital* (Basil Blackwell, Oxford, 1985).
6. David Harvey, "Possible Urban Worlds: A Review Essay," *City and Community* (March 2004): 83–89.
7. Lefebvre, *L'idéologie structuraliste* (Anthropos, Paris, 1975), p. 70. The bulk of the essays in this collection first appeared four years earlier in *Au-delà du structuralisme*.
8. *L'idéologie structuraliste,* p. 11. "Today," Lefebvre said, in his 1975 preface, "where the structuralists see themselves as the object of convergent attacks, the sole regret of this author is to not have taken his polemic further and pushed it more forcefully."
9. Ironically, this schema is almost proto*regulationist* in design, a school whose intellectual roots are often associated with Althusser, Lefebvre's antihumanist archenemy. Lefebvre's francophone interpreters, people like Jacques Guigou and Remi Hess, talk of his post-'68 "Althusserian *dérive*." The subtitle alone of *The Survival of Capitalism* speaks volumes: "The Reproduction of Relations of Production." The duo likewise claims Lefebvre's 1970s œuvre contained analysis that could be construed as "institutional," reflecting society's (and Lefebvre's own?) loss of revolutionary momentum. See Remi Hess, "Préface à la troisième édition de '*La survie du capitalisme*' "; and Jacques Guigou, "La place d'Henri Lefebvre dans le Collège invisible, d'une critique des superstructures à l'analyse institutionelle." For more on Althusser and the reproduction of capitalist social relations, see my *Metromarxism,* pp. 114–18.
10. With twenty-years hindsight, David Harvey confirmed what Lefebvre here only hints: the passage "from managerialism to entrepreneurialism" in urban and global governance. See "From Managerialism to Entrepreneurialism: The Transformation in Urban Governance in Late

Capitalism," *Geografiska Annaler* 71B (1989): 3–17. Since the mid-1970s, social democratic managerialism, whose mainstay was an interventionist state concerned about redistributive justice, has steadily dissolved into a bullish entrepreneurialism. Therein, "lean" government divests from collective consumption obligations, public housing, health care, and education, and enters into so-called public–private partnerships. The corporate sector has had a jamboree, cashing in on welfare handouts for private speculation. What was meant to "tickle down" to the urban poor has invariably, Harvey stresses, flowed out into pockets of the already rich.

11. Lefebvre, *Vers le cybernanthrope: contre les technocrates* (Denoël, Paris, 1971), p. 194.
12. *Vers le cybernanthrope,* pp. 196–98. This little gem of a text, which screams out for close reading and English translation, exhibits some of Lefebvre's liveliest prose since *La Somme et le Reste.*
13. Lefebvre would develop this idea in *The Production of Space,* published four years on. There, he'd counterpoise jargon with argot, representations of space with spaces of representation. Argot's power is a power of ribald words, recalling that it is dangerous to speak: sometimes too much, sometimes too little. For one of the best scholarly treatises on argot, see *Les princes du jargon* (Gallimard, Paris, 1994), written by Guy Debord's widow Alice Becker-Ho. Of course, the literary giant of argot, of the disorderly mind the street embodies, is a scribe Lefebvre (and Debord) both admired: Louis-Ferdinand Céline, especially his 1936 masterpiece *Mort à crédit* [Death on Credit].
14. *Vers le cybernanthrope,* p. 213.
15. Ibid., p. 212.
16. Ibid., p. 213.
17. Lefebvre, *La Proclamation de la Commune* (Gallimard, Paris, 1965), pp. 20–21.
18. Ibid., p. 26.
19. Ibid., p. 32.
20. Ibid., p. 39.
21. Ibid., p. 40. Lefebvre's interpretation of the Commune led to blows with Guy Debord and the Situationists, who accused their former comrade of pilfering Situ ideas on 1871. "A certain influence has been attributed to Lefebvre," Debord wrote in a pamphlet called "The Beginning of an Era" (1969), "for the SI's radical theses that he surreptitiously copied, but he reserved the truth of that critique for the past, even though it was born out of the *present*"; *Situationist International Anthology* (Bureau of Public Secrets, Berkeley, 1989), pp. 227–28. Debord reckoned Lefebvre's take on the 1871 Paris Commune was lifted from SI's "Theses on the Commune" (1962). "This was a delicate subject," Lefebvre later recalled in a 1987 interview. "I was close to the Situationists. … And then we had a quarrel that got worse and worse in conditions I don't understand too well myself. … I had this idea about the Commune as a festival, and I threw it into debate, after consulting an unpublished document about the Commune that is at

the Feltrinelli Institute in Milan. I worked for weeks at the Institute; I found unpublished documentation. I used it, and that's completely my right." "Listen," insisted Lefebvre, "I don't care at all about these accusations of plagiarism. And I never took the time to read what they wrote about the Commune in their journal. I know that I was dragged through the mud." Curiously, Lefebvre thanks Debord in *La Proclamation de la Commune* (p. 11, footnote 1), for his friendship and support "in the course of fecund and cordial discussions." But in a typesetting howler (or a Lefebvre practical joke?), Debord is cited as M. Guy Debud!

22. Régis Debray, *Revolution in the Revolution? Armed Struggle and Political Struggle in Latin America* (Monthly Review Press, New York, 1967), pp. 76–77.

23. This is a crucial passage in *The Urban Revolution.* Alas, Minnesota University Press's English translation has deflected Lefebvre's original meaning.

Chapter 6

1. *Espace et société,* which Lefebvre launched with Anatole Kopp and Anthropos's blessing, was formative in his spatial turn. Between 1970 and 1980, the journal was a mouthpiece for New Left thinking on cities, space, and politics, as well as an outlet for a new breed of Young Turk critical sociologists, economists, and political scientists. (The sociologist Manuel Castells was a member of the *Espace et société* collective.) Issue number 1 (November 1970) was inaugurated with Lefebvre's pioneering "Réflexions sur la politique de l'espace," an agenda-setting manifesto. "I thus repeat," Lefebvre wrote, "there is a politics of space because space is political." The article was reprinted in Lefebvre's *Espace et politique;* an English translation appeared in the radical geography journal *Antipode, Espace et société*'s nearest Anglo-Saxon counterpart, spearheaded in the United States by the geographer Dick Peet. See Henri Lefebvre, "Reflections on the Politics of Space," *Antipode* 8, no. 2 (1976): 30–37; the piece also featured in Peet's handy (and still valuable) edited collection *Radical Geography* (Maaroufa Press, New York, 1977).
2. See Remi Hess, "Henri Lefebvre et la pensée de l'espace," Avant-Propos à la quatrième édition française de *La production de l'espace* (Anthropos, Paris, 2000), p. xiv.
3. Ibid., pp. xv–xvi.
4. Donald Nicholson-Smith, the translator of *The Production of Space,* passed this information on to me in an e-mail exchange, April 21, 2005.
5. Guy Debord, the other Hegelian Marxist theorist, was equally nowhere on "respectable" Anglo-American theoretical curricula. *The Society of the Spectacle,* pirated by Fredy Perlman's anarchist Black and Red Books in Detroit and later by London's Rebel Press, was exclusively fringe-militant nourishment.

6. Manuel Castells, *The Urban Question* (Edward Arnold, London 1977), p. 87. Emphasis in original.
7. In a personal communication, Marshall Berman told me that throughout the 1970s he tried to talk various publishers into translating Lefebvre. "Hopeless!" was how Berman described it.
8. Cited in Gailia Burgel et al., "An Interview with Henri Lefebvre," *Environment and Planning D: Society and Space* 5 (1987): 27–38. For more on Castells's urbanism and Althusserian inflections, see *Metromarxism,* chap. 6.
9. The page references that follow refer to Donald Nicholson-Smith's English translation (e.g., Basil Blackwell, Oxford, 1991).
10. Karl Marx, "Critique of Hegel's Philosophy of Right. Introduction," in *Karl Marx—Early Writings* (Penguin, Harmondsworth, 1974), p. 251.
11. Marx, *Capital—Volume 1* (Penguin, Harmondsworth, 1976), chap. 1, pp. 165–66.
12. Spaces of representation express the realm Lefebvre deemed *habiter* in *la droit à la ville* and *La révolution urbaine*. Heidegger's influence is obvious here, as is *The Poetics of Space* (1957) of French philosopher Gaston Bachelard (1884–1962). "With Bachelard's 'poetics of space,' " Lefebvre notes, "the contents of the House have an almost ontological dignity: drawers, chests and cabinets are not far removed from their natural analogues ... namely, the basic figures of nest, shell, corner, roundness. ... The House is as much cosmic as human. ... The shell, a secret and directly experienced space, for Bachelard epitomizes the virtues of human 'space' " (POS, p. 121).
13. In 1960, Kevin Lynch's classic *The Image of the City* (MIT Press, Cambridge, MA) first expounded how the realm of perception conditions a person's spatial practice in the city. Lefebvre hints at the class and social group applicability of this thesis, hooking it up with broader economic and political structures of power, those conditioning and affecting individual behavior and cognitive activity. This constitution and reproduction of daily life mimics Pierre Bourdieu's concept of "habitus," a "generative mechanism" whereby subjective dispositions tow an unconscious objective line. "Because," Bourdieu writes, with characteristic complexity, "habitus has an endless capacity to engender products—thoughts, perceptions, expressions, actions—whose limits are set by the historically and socially situated conditions of its production, the conditioned and conditional freedom it secures is as remote from a creation of unpredictable novelty as it is a simple mechanical reproduction of the initial conditionings"; see Pierre Bourdieu, *Outline of a Theory of Practice* (Cambridge University Press, Cambridge, 1977), p. 95. David Harvey has suggested habitus is a "very striking depiction" of the constraints to the power of the lived over the conceived; see "Flexible Accumulation through Urbanization," *Antipode* 19 (1987): 268.
14. Octavio Paz, *Conjunctions and Disjunctions,* trans. Helen Lane (Arcade Publishing, New York, 1990), p. 115.

15. Lefebvre, *Le manifeste différentialiste* (Gallimard, Paris, 1970), p. 45.
16. Ibid., p. 156.
17. Ibid., p. 131.
18. Ibid., pp. 49–50. Emphasis in original.
19. See ibid., chap. IV, "Contre l'in-différence."
20. Nietzsche proclaimed that "God is dead" for the first time in *Gay Science* (1882, Section 125). "God remains dead," he followed up. "And we have killed him. How shall we, the murderers of all murderers, comfort ourselves?" *The Portable Nietzsche* (Viking Press, New York, 1967), p. 95.
21. Friedrich Nietzsche, "The Birth of Tragedy," in *Basic Writings of Nietzsche* (Random House, New York, 1966), p. 37.
22. Lefebvre, "Préface au quatrième édition de *La production de l'espace*," pp. XXVI–XXVII.

Chapter 7

1. Lefebvre, *The Survival of Capitalism* (Alison and Busby, London, 1976), p. 21. Emphasis in original.
2. A succinct, essay version of Lefebvre's vision of the capitalist state, with specific reference to the French experience, can be found in his "Comments on a New State Form," *Antipode* 33 (2003): 769–82 (translated by Victoria Johnson and Neil Brenner). This citation is found on page 773. The text is accompanied by Brenner's user-friendly preface "State Theory in the Political Conjuncture: Henri Lefebvre's 'Comments on a New State Form' "; *Antipode* 33 (2003): 783–808.
3. Lefebvre, "Comments on a New State Form," p. 774.
4. Ibid., p. 774.
5. Lefebvre in "Une Interview d'Henri Lefebvre," *Autogestion et socialisme* 33/34 (1976): 121–22.
6. Lefebvre, "Comments on a New State Form," p. 774.
7. Ibid., p. 773.
8. Ibid., p. 772.
9. Ibid., p. 779.
10. Richard O'Brien, *Global Financial Integration: The End of Geography* (Council on Foreign Relations Press, New York, 1991).
11. Michael Hardt and Antonio Negri, *Empire* (Harvard University Press, Cambridge, MA, 2000), p. 23. Parenthetical references in text refer to this edition.
12. Karl Marx, *Critique of the Gotha Program* (Progress Publishers, Moscow, 1978), p. 20.
13. Ibid., p. 21 (emphasis in original).
14. Lefebvre, *The Sociology of Marx* (Penguin, Harmondsworth, 1968), p. 184.
15. Ibid., pp. 182–83.
16. Lefebvre, "Henri Lefebvre ouvre le débat sur la théorie de l'autogestion," *Autogestion et socialisme* 1 (1966): 60.

17. Lefebvre, *Autogestion et socialisme,* pp. 60–61. In *De l'Etat 4: les contradictions de l'état moderne* (Collection 10/18, Paris, 1978), Lefebvre explicitly mobilizes dialectical wisdom to analyze the "modern" state. The state here cannot be considered as eternal, as per Hegel, he says, and it cannot be abolished directly, as per the anarchists. Rather, the state needs to be "subordinated" to society and "reabsorbed" within society. Hence Lefebvre's critique of the state ends up offering a new definition of socialism, which is neither a Leninist dictatorship of the proletariat nor a multitude rising without institutional (and place) mediation.
18. Lefebvre, *de l'État 3: le mode de production étatique* (Collection 10/18, Paris, 1977), p. 151.
19. Marx, *Capital—Volume 1* (Penguin, Harmondsworth, 1976), p. 287, footnote 7.
20. Ibid., p. 993.
21. Hardt and Negri, *Empire,* pp. 190, 209, 211.
22. In fact, Hardt and Negri's smooth, unruffled globe has a distinctive Kantian feel to it, a realm of an almost ideal space, a transcendental and essentially ungraspable structure—a feature Lefebvre denounces with gusto (cf. POS, p. 2). Space here could easily be construed as a purely *classifying* phenomenon, belonging to the a priori realm of the multitude's consciousness, separable from the empirical sphere.
23. Lefebvre, *Everyday Life in the Modern World,* p. 14.
24. José Bové and François Dufour, *The World Is Not for Sale: Farmers against Junk Food* (Verso, London, 2001), pp. 4–5.
25. Ibid., p. 13.
26. Lefebvre's old mate Guy Debord made a similar point a couple of years before Lefebvre's death. In his short autobiography *Panégyrique* (1989) ([Verso, London, 1991], pp. 47–48), Debord wrote, "Nearly all alcohol, and all beers … have today entirely lost their taste, first on the world market, then locally, with the disappearance or economic re-education of social classes who were long independent of large industrial production, and so too by the play of various state rules who from now on almost prohibit all that isn't manufactured industrially. … In the memory of a drunkard one never imagined that they would see drinks in the world disappear before the drinker."
27. Bové and Dufour, *The World Is Not for Sale,* p. 30.
28. Lefebvre, "Henri Lefebvre ouvre le débat sur la théorie de l'autogestion," *Autogestion et socialisme* 1 (1966): 62.
29. Lefebvre, "Comments on a New State Form," p. 780.
30. Lefebvre, "Une interview d'Henri Lefebvre," p. 123.
31. Lefebvre, *The Survival of Capitalism,* p. 124.

Chapter 8

1. Lefebvre, *La Somme et le Reste—Tome II,* p. 456.

2. Henri Lefebvre and Norbert Guterman, *La Conscience Mystifiée* (Éditions Syllepse, Paris, 1999), p. 66.
3. Ibid., p. 66.
4. *La Somme et le Reste—Tome II,* pp. 453–55. Cf. *Conversation avec Henri Lefebvre,* pp. 43–44.
5. Ibid., pp. 453–54.
6. Ibid., p. 453.
7. Hegel, *Phenomenology of Spirit* (Oxford University Press, Oxford, 1977), p. 124. Emphasis in original.
8. Ibid., p. 126.
9. Jean Wahl, *Le Malheur de la Conscience dans la Philosophie de Hegel* (Gérard Monfort, Paris, 1989), p. 9.
10. See Karl Marx, "Critique de la dialectique hégélienne," trans. Lefebvre and Guterman *Avant-Poste,* juin 1933, pp. 32–39. With Norbert Guterman, Lefebvre translated Hegel, Marx, and Lenin (on Hegel) and presented each in volumes called *Morceaux choisis de Hegel, Morceaux choisis de Marx,* and *Cahiers de Lénine sur la dialectique de Hegel,* respectively.
11. Marx, "Economic and Philosophical Manuscripts," in *Karl Marx—Early Writings* (Penguin, Harmondsworth, 1974), p. 386.
12. Ibid., p. 382.
13. Twenty-odd years later Lefebvre came clean: *La Somme et le Reste—Tome II* (p. 452) said that Lukács's confounding of economic and political consciousness, holy writ for the Third International, was "pernicious" for workers' struggle. The Third International (1919–43) became the sequel to the initial Marx and Engel's inspired "First International Workingmen's Association," the international communist movement's attempt to "unite workers of the world." After the collapse of the "Second International" (1889–1914) as World War I broke out, the Third International, or Comintern, was established by Lenin in Moscow in March 1919 following the October 1917 revolution. The statutes that followed, until the renegade Trotsky founded the Fourth International in 1938, proclaimed a "World Union of Socialist Soviet Republics." In *The Survival of Capitalism* (p. 50), Lefebvre wrote, "The lifespan of the Third International, a revolutionary organization transformed into a Stalinist institution, saw political thought and theoretical research completely crushed."
14. Georg Lukács, *History and Class Consciousness* (Massachusetts Institute of Technology Press, Cambridge, 1971), p. 86.
15. For a more detailed summary, see my *Metromarxism* (Routledge, New York, 2002), pp. 26–29.
16. Lukács, *History and Class Consciousness,* p. 197.
17. There's a curious sense in which Lefebvre's critique of Lukács's "concrete universal" equally applies to the thesis of "the multitude" espoused by Michael Hardt and Toni Negri in *Empire*. The act of faith through which Lukács posited the proletariat as the "identical subject-object of history" mirrors to a tee Hardt and Negri's "faith" in the multitude: "The concrete universal," Hardt and Negri say, "is what allows the multitude to pass from

place to place and make its place its own" (p. 362). Interestingly, Lefebvre's old Hungarian sparing partner is one thinker Hardt and Negri *don't* actually mention.

18. *La Conscience Mystifiée,* p. 71.
19. Ibid., p. 81.
20. Ibid., p. 70.
21. Ibid., p. 70.
22. Throughout the 1930s, Lefebvre worked on a companion volume to *La Conscience Mystifiée* called *La Conscience Privée;* the two texts were to be part and parcel of a larger project titled *Sciences des Idéologies.* Alas, the dialectical counterpart to the 1936 book was never completed. For years, Lefebvre's draft manuscript, which he dedicated to Norbert Guterman—"reduced to silence the other side of the Atlantic"—was thought destroyed. But it resurfaced in the 1950s, and republished editions of *La Conscience Mystifiée* contain the fated text as an annex. A fascinating typescript version of *La Conscience Privée,* bearing Lefebvre's scribbled annotations, can be glimpsed among Guterman's archives at New York's Columbia University.
23. "Economic and Philosophical Manuscripts," pp. 351–52. Emphasis in original.
24. Fyodor Dostoevsky, "The Grand Inquisitor," in *Notes from Underground and the Grand Inquisitor,* ed. and trans. Ralph Matlaw (E.P. Dutton Books, New York, 1960), p. 125.
25. *La Conscience Mystifiée,* p. 71.
26. *La Somme et le Reste—Tome II,* pp. 557–58.
27. *La Conscience Mystifiée,* p. 161.
28. Lefebvre, *Nietzsche* (Éditions Syllepse, Paris, 2003). As well as his *Nietzsche* text, Lefebvre wrote a "balance-sheet" in Germany, detailing five years of Hitler since the burning of the Reichstag, called *Hitler au pouvoir, bilan de cinq années de fascisme en Allemagne* (1938), even analyzing selections of *Mein Kampf* en route.
29. *La Somme et le Reste—Tome II,* pp. 464–65.
30. Ibid., p. 463.
31. Ibid., p. 465.
32. The state as a "cold monster" passage Lefebvre cites often; cf. *De l'État, La Conscience Mystifiée, Hegel-Marx-Nietzsche,* and *La Somme et le Reste.* For the source, see Friedrich Nietzsche, "Thus Spoke Zarathustra," in *The Portable Nietzsche,* ed. Walter Kaufmann (Viking Press, New York, 1954), pp. 162–63.
33. Nietzsche, "Thus Spoke Zarathustra," p. 163.
34. Karl Marx, *Capital—Volume 1,* pp. 340–44.
35. Friedrich Nietzsche, *The Will to Power* (Vintage Books, New York, 1968).
36. Lefebvre, *Nietzsche,* p. 117.
37. Ibid., p. 116.
38. Ibid., p. 117.

39. Nietzsche, *The Will to Power,* p. 17.
40. *Nietzsche,* p. 117; *La Somme et le Reste—Tome II,* p. 473.

Afterword

1. Henri Lefebvre, *Dialectical Materialism* (Jonathan Cape, London, 1968), p. 164.
2. Lefebvre, *Critique of Everyday Life—Volume 1,* pp. 66–67.
3. Lefebvre, *Conversation avec Henri Lefebvre,* p. 19.
4. Francis Fukuyama, *The End of History and the Last Man* (Free Press, New York, 1991).
5. Speech to Congress (March 6, 1991) in the wake of the first Gulf War. Poppy Bush's words are worth recalling: "Tonight in Iraq, Saddam walks amidst ruin. His war machine is crushed. His ability to threaten mass destruction is itself destroyed."
6. See the debate over "The Neoconservative Moment" and "Democratic Realism" that ensued in the pages of the *National Interest* (summer and fall issues, 2004) between liberal bourgeois Fukuyama and crackpot neocon Charles Krauthammer.
7. Krauthammer, "In Defense of Democratic Realism," *National Interest,* fall 2004, p. 15. Emphasis added.
8. *Dialectical Materialism,* p. 113.
9. See *Conversation avec Henri Lefebvre,* p. 23.
10. Franz Kafka, *The Castle* (Minerva, London, 1992), pp. 59–60.
11. Milan Kundera, *Le Rideau* (Gallimard, Paris, 2005), p. 152.

INDEX

L